This first biography of Lawrence Johnston, the creator of one of the most beautiful gardens in England, drawing upon new research carried out during the past eight years and illustrated by many previously unpublished images, sets out to answer the question *Who was Lawrence Johnston?* This remarkable man was an American born in Paris, France in 1871 who came to England in 1893 to study at Trinity College, Cambridge. He became a naturalized British citizen in 1900 and then served as a Private with the Northumberland Hussars in the Boer War in South Africa. By 1904 he had been elected a Fellow of the Royal Horticultural Society and came, in 1907, to Hidcote with his mother, Gertrude, a New York socialite. The next seven years were to see the creation of the central features of the garden at Hidcote and Lawrence's plantsmanship was recognised in 1911 by an award of merit from the R.H.S. for the Hidcote strain of *Primula pulverulenta.* After serving as a Major in World War I, Lawrence completed the garden and in the 1920s went on and sponsored plant hunting expeditions looking for the most attractive species for both Hidcote and the second garden that Lawrence created at Serre de la Madone, just inland from Menton on the Mediterranean coast of France. In 1948, Lawrence gave Hidcote to the National Trust as the first garden of national importance to be saved under a Gardens Fund launched jointly by the Trust and the R.H.S. As Lord Aberconway, the President of the R.H.S., said when presenting the Veitch Gold Medal to Lawrence *There has been no more beautiful formal garden laid out since the time of the old Palace of Versailles than that designed on quite a small scale, but with exquisite artistry, by Major Lawrence Johnston at Hidcote.* Lawrence, a generous man with many friends, died at Serre de la Madone in 1958 and is buried at Mickleton, a mile from Hidcote.

Lawrence Waterbury Johnston (1871—1958)

Lawrence Johnston

The Creator of Hidcote

Graham S. Pearson

National Trust, Hidcote, Chipping Campden, GL55 6LR

To Susan

for her encouragement, unfailing support

and assistance over 50 years

First published in the United Kingdom in 2010 by Hidcote Books, Hidcote Bartrim, Chipping Campden, GL55 6LR.

ISBN 978-0-9565051-0-1

British Library Cataloguing in Publication Data. A catalogue record for this book is available from the British Library.

Printed in England by the Vale Press Ltd., Willersey

Mixed Sources
Product group from well-managed forests, controlled sources and recycled wood or fiber
FSC www.fsc.org Cert no. CU-COC-807273-AQ
© 1996 Forest Stewardship Council

CONTENTS

FOREWORD

Lawrence Johnston is today well known as the creator of one of the most beautiful gardens in England – Hidcote Manor Garden. However, very little was known about Lawrence Johnston and how he came to create this garden as no records were provided when he gave Hidcote to the National Trust. Over the years information has come to light – notably through the research carried out in the late 1990s by Katie Fretwell, the then National Trust gardens historian, which focused primarily on the garden at Hidcote. However, some inaccurate accounts have been published and, once published, these inaccuracies tend to be recycled. Based on the examination of primary sources, this book builds on the earlier research and uses new information to provide an accurate account of this remarkable man, the family from which he came and the gardens that he created.

My research was triggered by the appearance in the winter of 2002 of a slim notebook covering the years from 1925 to 1928 and two engagement diaries for 1929 and 1932. These contain Lawrence Johnston's notes and engagements covering those years.

Lawrence Johnston's Notebook and Diaries for 1929 & 1932

I have subsequently visited several places as well as archives and libraries in the British Isles, France and the United States to seek and uncover new information about Lawrence, his family and his gardens. Many friends have helped in this work and have encouraged this search for new information.

In regard to his family, I have found, for example, that his parents divorced in the early 1880s and his father, Elliott, then aged 59, married a young girl of 18 and went to live on the Eastern Shore of Virginia until he died in 1901. We now know that his mother, Gertrude, took good care to look after her son financially by making two wills on the same day in London in 1924 – one for her estate in the United Kingdom and the other in the United States of America – so as to minimise death duties. And as to Lawrence himself, he was elected a fellow of the Royal Horticultural Society in 1904 three years before he came to Hidcote and he was borrowing books on gardening and garden design – including Thomas Mawson's *The Art and Craft of Garden Making* – from 1905 onwards. Furthermore, his plantsmanship was recognised as early as June 1911 less than four years after arriving at Hidcote when he received an Award of Merit from the R.H.S. for the Hidcote strain of *Primula pulverulenta* (also known as *Candelabra primula*). He was an early member of the Garden Society and had many friends amongst the landowners of Britain who, like him, were enthusiastic gardeners. His diaries show that he, like his mother, was a sociable person.

The new information enables us to gain a far better appreciation of Lawrence Johnston and of his creation of Hidcote. This is being used by the National Trust in their Conservation Plan for the garden that aims to recreate Hidcote as it was in its heyday in the 1930s.

ONE

AN AMERICAN FAMILY IN THE 1800s

Lawrence Waterbury Johnston was born on 12 October 1871 in Paris, France. He was the first child for Elliott Johnston, then aged 45, and Gertrude Cleveland Johnston (née Waterbury) who was at that time aged 25, some 20 years younger than her husband. His parents had married a year earlier in Westchester, New York City on 20 October 1870. In this chapter, we first explore his father's family and then his mother's family, Interestingly, Lawrence Johnston was related to two First Ladies to US Presidents, albeit almost a century apart, one on each side of his family.

Lawrence Johnston's father's family

Elliott Johnston was born on 1 May 1826 in Baltimore, Maryland into a family of stockbrokers and bankers. His parents, Thomas Donaldson Johnston and Anna Maria Elizabeth Elliott, had married a year earlier and Elliott was the first child of a large family of nine. He grew up in Baltimore, a major seaport situated on an arm of the Chesapeake Bay that extends some 200 miles into the east coast of the United States. The Chesapeake Bay has long been historically important as it was in 1607 at Jamestown, that the first permanent English settlement in America was established.

Elliott's father, Thomas, was for many years a partner in the banking house of Lee & Johnston and died suddenly from *an attack of apoplexy* or a stroke, aged 50, on 30 June 1851 *leaving a large and affectionate family, an extended circle of relatives and friends, and all the community, to regret his death.* These were good years in which to be engaged in banking as Baltimore is the largest city in Maryland and was then a major shipping and manufacturing centre which was closer than any other seaport on the east coast to the major Midwestern markets of America.

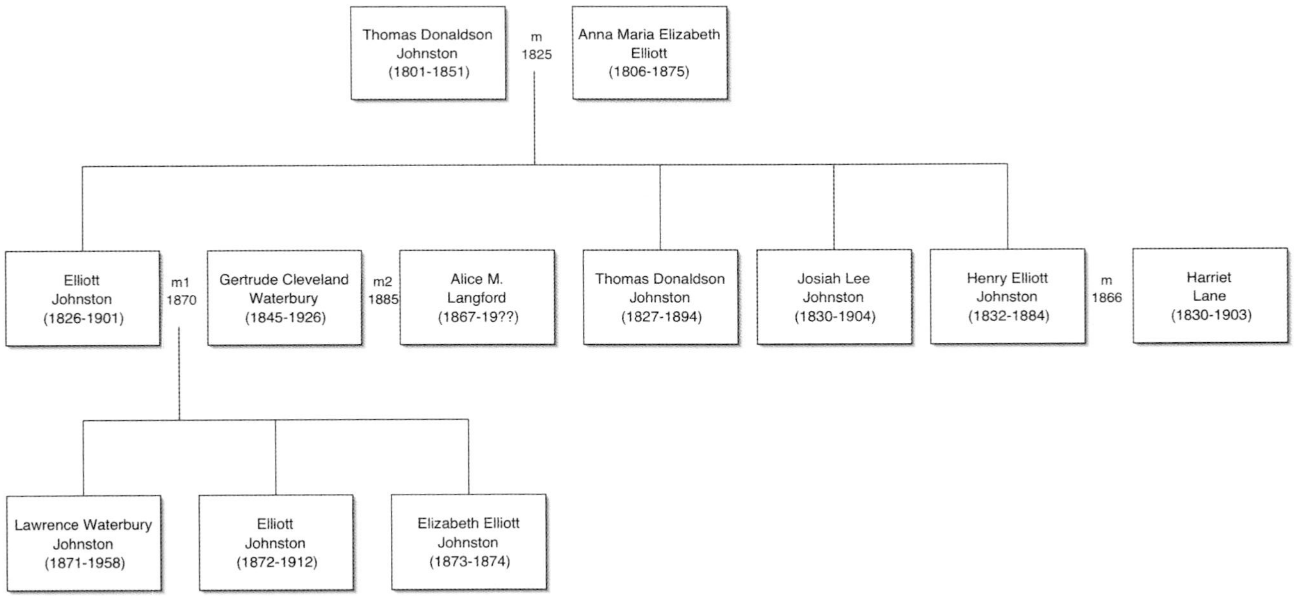

The Johnston Family Tree

The location of Baltimore in relation to New York City and to Washington, D.C. is shown on the map below.

Eastern United States map showing Baltimore & New York

An interesting insight into the origins of the Johnston family of Baltimore is provided by the biography of one of Elliott's brothers, Josiah Lee, who went to Harvard and graduated in 1849. This says the Johnston family were originally Scottish and moved to Ireland settling in the County of Fermanagh in 1457. A son, John, was a volunteer in the Enniskillen Horse and was at the siege of Londonderry and fought in the Battle of the Boyne under the Prince of Orange, later King William III. John after the war settled in Dublin and had a son, Samuel, who became an attorney. He in turn had a son, Samuel, who came to America in 1753 and had seven children. The youngest, George, married Margaret Skirvan Wilson and it was their son Thomas Donaldson Johnston born 23 April 1801 who married Elizabeth Elliott, only daughter of Hartman Elliott, formerly of Hesse Cassel, Germany

Josiah's biography also says he received his early education at private schools and under private tuition before he went to Harvard in 1847 and graduated with the class of 1849. He became engaged in banking in Baltimore in 1853 and for the following thirty years conducted an extensive and profitable business, retiring in March 1884. Together with another of Elliott's brothers, Henry, he continued the family business after their father's death. In the 1850s this traded as the *Johnston (Josiah Lee and Henry E.) Brothers and Co,* Banking House at 198 Baltimore Street as a mis-spelt entry in the Baltimore Directory for 1855-56 showed.

JOHNSON (Josiah Lee & Henry E.,) BRO. & CO., Banking house, 198 Baltimore
Johnson (Josiah Lee & Henry E.) Bro. & Co, h 17 s Howard

Baltimore Directory, 1855-56

The family is shown in Baltimore Directories in the 1830s, 1840s and 1850s as living at 17 South Howard Street near Cider Alley. They were a wealthy family as the 1860 census shows the value of real estate and personal estate for the two sons – Josiah Lee and Henry – and their mother amounted to some $375,000 equivalent to almost $9 million today.

Lawrence's father, Elliott, as the eldest son seems to have followed the tradition in many English families of the first son serving in the armed forces. He was appointed as a Midshipman in the United States Navy on 17 September 1841 at the age of 15. There are sparse records at the U.S. Naval Academy, Annapolis, Maryland – a few miles south of Baltimore – showing that he was first posted to the *Brandywine*, a 44 gun frigate, with 467 men and officers, which in 1843 took him round the Cape of Good Hope to Bombay and then on to the Orient to Macao, Manila and Hong Kong. The *Brandywine* returned via Hawaii to the west coast of South America before rounding Cape Horn.

U.S. Frigate Brandywine *ca. 1831*

He then in 1844 was transferred to the *Columbus*, a 74-gun ship of the line which sailed in 1845 for Canton, China and then went on to Japan. At a later stage, he was on the *Portsmouth*, a wooden sloop of war, which sailed in the Pacific off the coasts of California and Mexico. Then, at the age of 23, Elliott resigned from the US Navy on 16 August 1849. During the next ten years, he travelled widely as his obituary in 1901 recorded that *He was of a roving disposition and prior to the Civil War he visited many foreign countries.*

When the Civil War began in 1861, Elliott was in Baltimore and was among the first to go south to offer his services to the Confederate Forces. Although the State of Maryland did not join the Confederate States, it is known that many in Baltimore had Confederate sympathies which led to the first conflict of the Civil War being the Baltimore riot of April 1861. Elliott initially served in December that year as a volunteer Aide de Camp (A.D.C.) on the staff of General R. B. Garnett. Then in June 1862 Elliott wrote a letter to Lt. Colonel Chilton, the Assistant Adjutant General of the Confederate Forces, requesting that he be commissioned and given authority to raise a battery of artillery to serve under General 'Stonewall' Jackson. In this he describes himself:

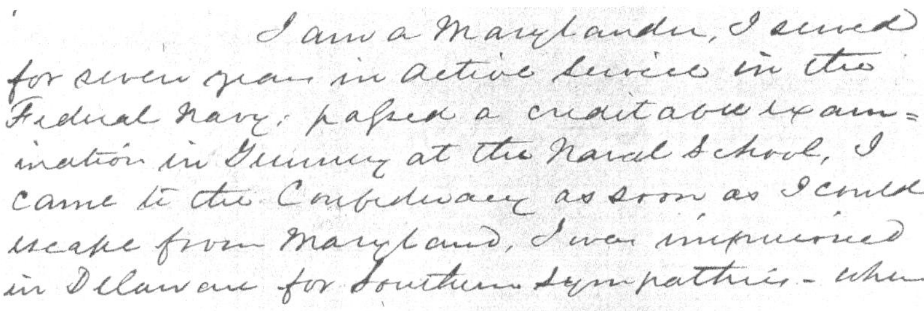
I am a Marylander, I served for seven years in active service in the Federal Navy; passed a creditable examination in Gunnery at the Naval School, I came to the Confederacy as soon as I could escape from Maryland, I was imprisoned in Delaware for Southern sympathies – when

Elliott Johnston letter to Lt. Colonel R. H. Chilton 3 June 1862

Elliott's letter shows that whilst a midshipman in the US Navy he passed an examination in gunnery at the Naval School and that when he came to enlist with the Confederate Forces he was imprisoned for his southern sympathies in Delaware, the State adjoining Maryland to the north. The same letter goes on to say that *I desire to state here, Col., that I have paid my own expenses through the war, never having cost the Confederacy a cent, I desire active service and beyond all I wish to be with the army when it goes into my native State.*

He enclosed a letter of commendation from General R. B. Garnett that stated *I feel it my duty to bring to your notice, my volunteer aid, Mr. Elliott Johnston of Maryland, who is very desirous of active service in the field. He has served with me since December last, and I have been much indebted to him for the prompt and intelligent manner with which he has discharged his duties. I beg leave to here introduce the mention made of him in my official report on the Battle of Kernstown.*

> *"Mr. Johnston accompanied me during the entire day, except when engaged carrying orders or rallying and encouraging troops. His conduct was characterized by the most dauntless bravery. His inspiring and chivalrous bearing challenged the admiration of all who saw him. I most earnestly recommend his promotion in the Confederate Service."*

I respectfully repeat the recommendation being confident that if it is granted it will add to the Southern cause an officer as brave and devoted to it as any now in commission.

Kernstown is a small town in the Shenandoah Valley of Virginia some 40 miles south-west of Harpers Ferry, on the Potomac River bordering

Maryland. The Battle of Kernstown on 23 March 1862 was the opening battle of Confederate Army Major General Thomas J. "Stonewall" Jackson's campaign through the Shenandoah Valley during the American Civil War.

Elliott's application succeeded as he was appointed A.D.C. with the rank of 1st. Lieut. to take effect from 24 March 1862. A letter from Elliott to his sister, Bessie, on 21 July 1862 provides an illuminating insight into the high spirits in the ranks of the Confederate troops:

> *I wish I could tell you how beautifully our Confederacy works. Every one is united; there is no bickering or quarrelling. "Little Mac" or the "Young Napoleon" or ... "The Great Ditch-Digger"* [referring to General George B. McClellan, the Commanding General of the Union Forces] *"pushed us to the war" so hard that the rebound threw him 37 miles from Richmond where he now is with his tail between his legs, like a whipped cur. ... I do not know how long the North can hold out in this war, but be assured we can never be subdued. The South is one bristling field of bayonets. ...*
>
> *That we suffer many privations is indisputable. Coffee and tea I have not seen for months. Of uniform we have scarcely a vestige. Many & many a wife & sister has cut up her calicos for shirts for the troops. We are almost shoeless – but you should see the troops – line after line – extending for miles of muskets that you can almost see to shave in – the most superb cannon presented to us by Mr. McClellan (52 in number), arms, ammunition, provision tents, etc, in uncountable numbers. It was a most glorious victory and only seemed incomplete by the negligence of a General officer – McClellan was in a bag with 16,000 of his men – but the string was not drawn in time & he got out. I suppose this will be a long war, but every day of it we are getting stronger, the North weaker. .... . We expect a struggle for Richmond again in a few weeks. It will not be as hard as before, but many a brave heart will be laid low – I have lost many friends in these battles.*
>
> *I will write often now, as frequent opportunities occur of sending letters. Give my love to all. I am glad to see you are true; be sure I never doubted you.*

His action in the Battle of Cedar Mountain that took place in Northern Virginia some 75 miles north-west of Richmond in August 1862 was reported in the newspapers.

> At length Lieut. Elliott Johnston, of the rebel General Garnett's staff, advanced down a slope bearing a white flag, which he waved vigorously, as if to attract our attention. Some of our officers at once went up to the Lieutenant, who informed them that by permission of Gen. Stonewall Jackson they might have till two o'clock to bury the dead.

The Daily Pittsburgh Gazette, 15 August 1862

Elliott continued to fight with the Confederate Forces and was injured, losing his left leg, in the Battle of Antietam (also known as the Battle of Sharpsburg) in September 1862 – this was the battle when more men were killed on a single day than any other day of the Civil War. It was the first major battle of the Civil War to take place on northern soil north of Harpers Ferry in Maryland, some 50 miles north-west of Washington DC. A note in the Baltimore *Sun* on 4 October 1862 under the heading of *Maryland wounded* about those in the Confederate Forces who had been wounded in the battle of Antietam records that *Elliott Johnston, formerly of the United States Navy and a native of this city, was so badly wounded that he was compelled to undergo amputation of the leg below the knee.*

Following the battle the Confederates retreated south to Sheperdstown across the Potomac in West Virginia and to the north west of Harpers Ferry. Shortly after being wounded, Elliott was captured at Sheperdstown by the Union Forces and released on parole on 30 September 1862. Parole was used extensively in the Civil War as it enabled the captors to avoid the responsibility and burden of guarding, feeding, clothing and providing medical aid to their prisoners of war. Prisoners who had been paroled were returned to their side but were prohibited from taking up arms until they had been formally exchanged with prisoners from the other side. Elliott was released from his parole along with 370 other prisoners on 18 February 1863.

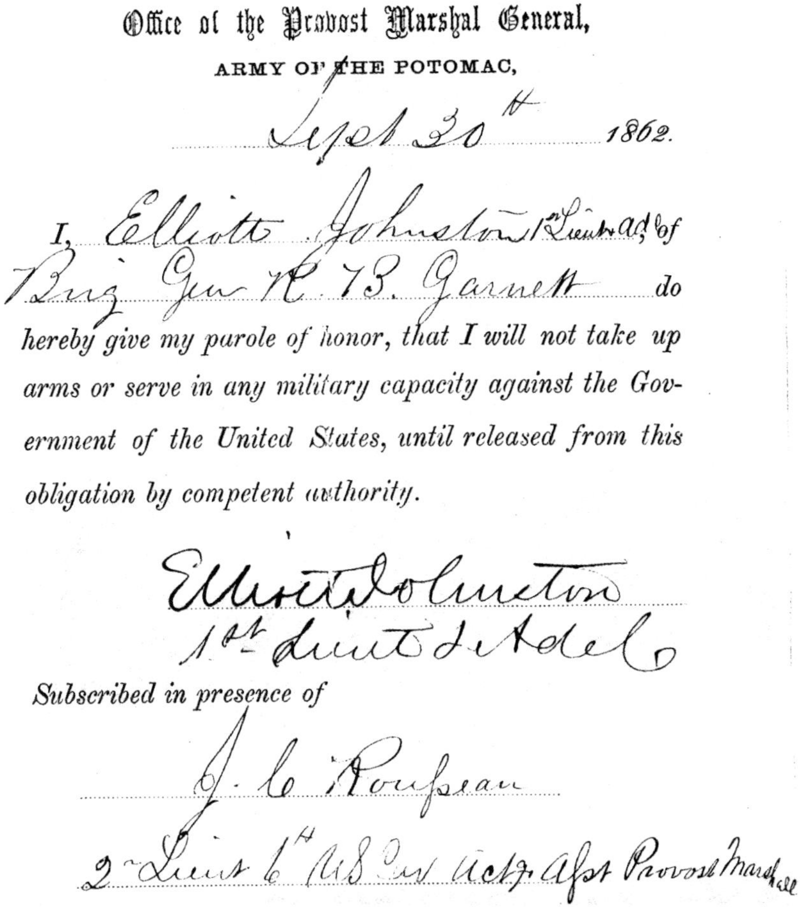

Office of the Provost Marshal General,

ARMY OF THE POTOMAC,

Sept 30th 1862.

I, Elliott Johnston 1st Lieut Adj of Brig Gen R. B. Garnett do hereby give my parole of honor, that I will not take up arms or serve in any military capacity against the Government of the United States, until released from this obligation by competent authority.

Elliott Johnston
1st Lieut & Adj

Subscribed in presence of

J. C. Rousseau
2nd Lieut 6th US Cav Actg Asst Provost Marshal

Elliott Johnston Parole 30 September 1862

After being granted parole, the next month, on 21 October, Elliott's brothers – probably Josiah Lee and Henry – were given permission to remove Elliott from Sheperdstown to Baltimore by General McClellan, in overall command of the Army of the Potomac, the major Union Army in the area near Washington, D, C. This permission is reproduced on the next page.

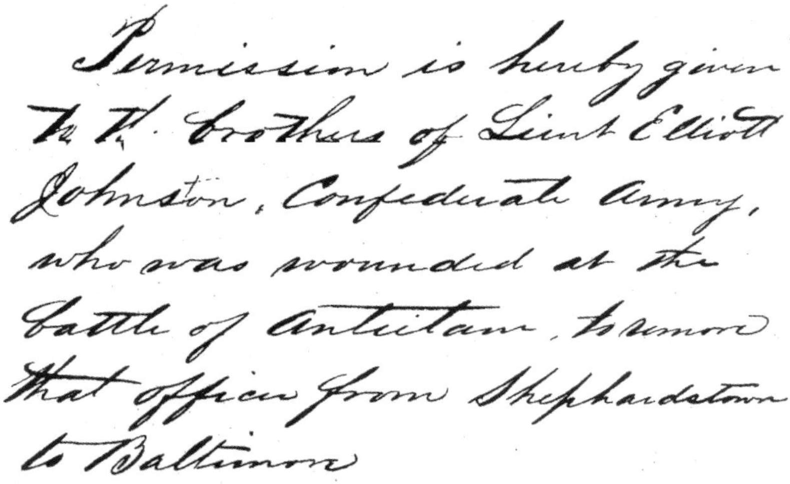
Permission is hereby given to the brothers of Lieut Elliott Johnston, Confederate Army, who was wounded at the battle of Antietam, to remove that officer from Shephardstown to Baltimore

21 October 1862 permission to remove Elliott Johnston to Baltimore

Elliott obtained an artificial leg and returned to the Confederate Forces and ordered to report to General Robert E. Lee on 28 May 1863. He served as an aide to General R. J. Ewell in the Gettysburg campaign in June and July 1863. He was promoted to Captain and Assistant Adjutant-General on 2 November 1863. It is clear that his left leg was causing him much pain and he was transferred from field service in November 1863. Matters did not improve in 1864 when Elliott was frequently on leave because of pain and he was frequently moved in an effort to find a suitable posting which did not involve field service or much exercise. By the end of the year he was seeking leave of absence with permission to visit Europe to seek a replacement artificial leg.

Although Elliott's letter to his sister early in the war made it clear that the Confederates had little in the way of uniforms, it is interesting that today there are three of Elliott's uniforms in the Maryland Historical Society. These all have State of Maryland Seal buttons and were tailor-made from superior quality wool-blend material. His captain's uniform had a blue-grey wool, single-breasted jacket which had a round, standing collar of cream-coloured ribbed cotton, which is also found on the lower sleeves. This jacket was worn during 28 battles and skirmishes.

Elliott Johnston wearing his captain's jacket

A letter sent on Elliott's behalf on 3 January 1865 to the Hon. James A. Seddon, the Secretary of War for the Confederate Government, sought permission for Elliott to travel to Europe to obtain a better artificial leg. This painted a glowing picture of him:

> *From a personal acquaintance with Capt. Johnston I can testify to his extraordinary worth; his disinterestedness in forsaking a home of wealth & luxury to espouse the Southern cause; his personal accomplishments intellectual & social; his gallantry & efficiency in the field strongly attested by all his superior officers; his persistent struggles under great personal suffering to fill up the measure of his duty, & his desire somehow to share the fortunes of an officer & not*

profit from meal in May, anxious to recruit his health & repair his shattered frame for future Service, he covets the opportunity to obtain the best mechanical and surgical aid which may be afforded in Europe; and expresses his willingness to bear his own expenses & to tax the Government as little as possible in the pursuit of this object.

A year after the Civil War ended in 1865, one of Elliott's younger brothers, Henry married Harriet Lane, the niece of President James Buchanan. As President Buchanan was a bachelor, he had asked Harriet, who lived with him as both of her parents had died by the time that she was eleven years old, to be his First Lady during his Presidency from 1857 to 1861. As such she is one of only two First Ladies not to have been the wife of the President.

Harriet Lane, US National Portrait Gallery

She had also travelled with her uncle to England where he was the US Ambassador to the Court of St. James from 1854 to 1855. Together with her uncle she moved in prominent social circles and dined along with her uncle with Queen Victoria and Prince Albert. During her uncle's presidency in 1860, Harriet was responsible for the social programme for the visit of the 18 year old Edward, the Prince of Wales to Washington, D.C. As the Prince had so enjoyed his visit to Washington, he invited Harriet Johnston, as she then was, to attend his coronation as King Edward VII on 9 August 1902, which she did. Harriet is one of nineteen ladies included in a book published in 1900 entitled *Famous American Belles of the Nineteenth Century* by Virginia Peacock.

> It is supposed that no American woman ever had more offers of marriage than Harriet Lane, and it is evident, from a letter written her by her uncle about this time, that suitors had already begun to present themselves. "I wish now to give you a caution," he wrote: "never allow your affections to become interested, or engage yourself to any person, without my previous advice. You ought never to marry any person who is not able to afford you a decent and immediate support. In my experience I have witnessed the long years of patient misery and dependence which fine women have endured from rushing precipitately into matrimonial connections without sufficient reflection. Look ahead and consider the future, and act wisely in this particular."
>
> *Harriet Lane in* Famous American Belles of the Nineteenth Century

Her marriage to Henry Johnston took place, with her uncle's blessing, on 11 January 1866 at Wheatland, the rural home and 22 acre estate of President Buchanan, in Lancaster County, Pennsylvania some 120 miles north of Washington, D.C. It is possible that Elliott was present as the report in the *New York Times* says that the brother and sisters of the groom were present. The honeymoon was spent in Cuba and her married life in Baltimore. Two years later, when President Buchanan died, Harriet inherited his estate at Wheatland which she then used as a summer retreat before selling it in 1884.

When Harriet died on 3 July 1903, she left to Lawrence Johnston, the eldest son of Elliott, *my son Harry's watch and chain* with a note that both the watch and chain belonged to his grandfather, Thomas Donaldson Johnston.

Five years after the end of the Civil War in 1865, Elliott married Gertrude Waterbury of Westchester, New York. Gertrude was the daughter of Lawrence Waterbury, owner of a rope-making factory in Brooklyn, New York.

Lawrence Johnston's mother's family

Gertrude Cleveland Waterbury was born on 21 December 1845 in Williamsburgh, New York City. Williamsburgh was annexed into Brooklyn in 1855; it is the northern part of Brooklyn looking west across the East River to Manhattan, which is shown as New York County below:

The Five Boroughs of New York City with Westchester County to the north

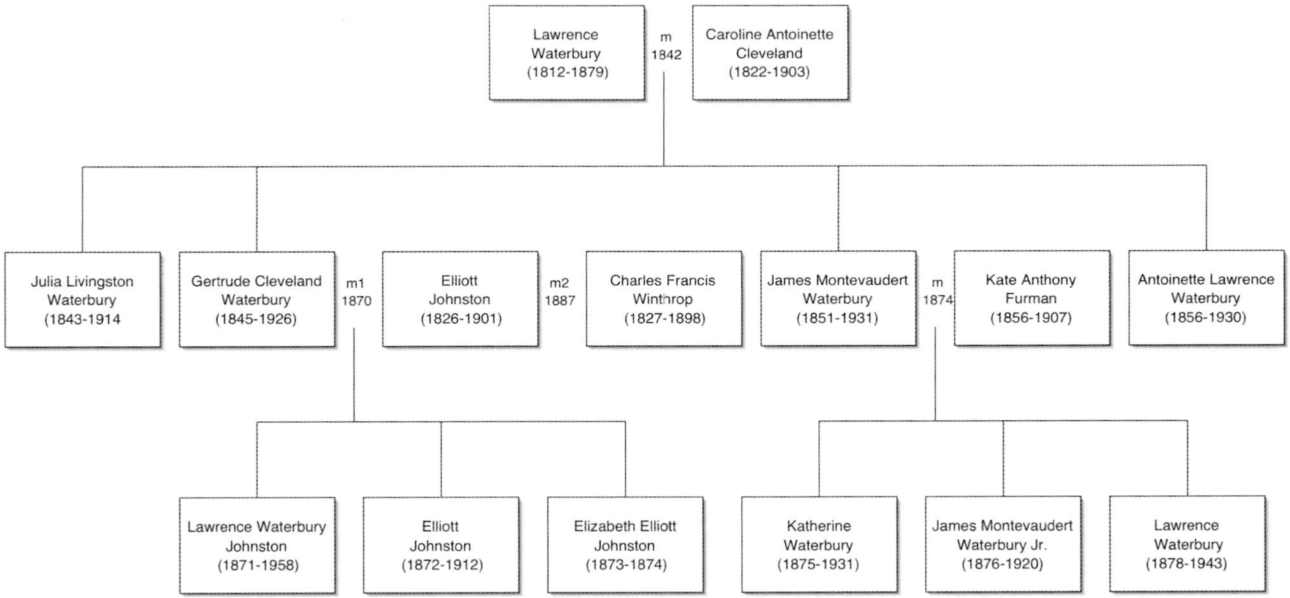

The Waterbury Family Tree

Gertrude's father, Lawrence Waterbury, aged 29, had married Caroline Antoinette Cleveland, aged 19, on 8 June 1842 in Williamsburgh. Gertrude, born 23 December 1845, was the second of their children. Her father was the youngest child of Noah Waterbury, who had been a leading figure in Williamsburgh. He had owned a ferry across to Manhattan and then established a rope making business as well as becoming involved in real estate. He created the City Bank and was its first President. The census of 1860 records that Noah then had real and personal estate valued at $170,000 (equivalent to $4 million today). When he died in 1862 the *New York Times* noted that *Out of respect for the deceased, the flags on the ferryboats and the shipping were raised at half-mast yesterday.*

His son, Lawrence, took over the rope making business and the census of 1860 shows that he was then living in Westchester, New York with real and personal estate of $150,000 (equivalent to $3.5 million today). There were six servants including a nurse from France. Ten years later, his value had almost doubled and there were seven servants. When Lawrence died in 1879 he was living at Throggs Neck in Westchester, New York, which is now part of the Bronx. His will of 8 March 1875 left his estate, then valued at $2 million (equivalent to $45 million today), to his four children:

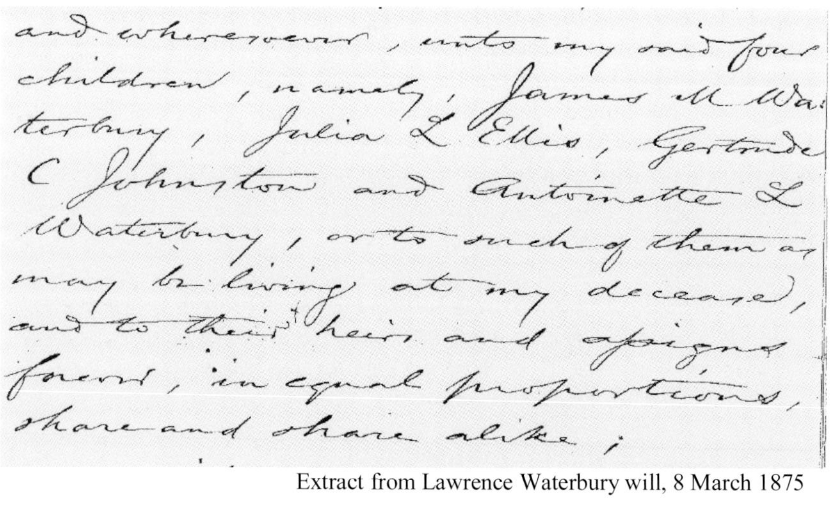
and wheresoever, unto my said four children, namely, James M Waterbury, Julia L Ellis, Gertrude C Johnston and Antoinette L Waterbury, or to such of them as may be living at my decease, and to their heirs and assigns forever, in equal proportions, share and share alike;

Extract from Lawrence Waterbury will, 8 March 1875

Lawrence was buried, as was his wife when she died some 24 years later in 1903, at St. Peter's Church, Westchester.

The Waterbury graves at St. Peter's Church, Westchester

Gertrude's brother, James, married Kate Anthony Furman on 22 April 1874; their children included James Montevaudert Jr. and Lawrence. Both sons became famous for their prowess at polo as they played for the United States in the international polo matches against England for the American Cup in 1902 and again in 1909. The American Cup was won by the England team in 1886 and although the US team lost in 1902 they won in 1909 in England. This win was reported in the *New York Times* on 5 July:

> *London, July 5. Lawrence Waterbury, J. M. Waterbury, Harry Payne Whitney, and Devereux Milburn, comprising the Meadow Brook polo team of Westbury, L. I., won the second and deciding match for the American Polo Cup here to-day.*
>
> *The score was 8 to 2.*
>
> *When Harry Payne Whitney as Captain of the team received from the Princess of Wales the trophy emblematic of the polo championship of*

the world no heartier congratulations were ever given to stranger[s] *by Englishmen.*

Never were compliments better deserved, as the challengers eclipsed everything in polo that had been seen here before. Experts all admit that England was beaten fairly and squarely and by a superior team.

One of the polo players, Gertrude's nephew, Lawrence Waterbury in 1900 married Miss Maude Ludlow Hall who was the aunt of Eleanor Roosevelt, the wife of President Franklin D. Roosevelt, who was U.S. President for four terms from 1933 to 1945. Consequently, Lawrence Johnston was related on his mother's side, albeit more distantly, to another First Lady of the United States.

Eleanor Roosevelt 1884-1962

Gertrude's younger sister, Antoinette, married John Pierrepont Edwards, the then Vice Consul of the British Consulate General in New York on 25 October 1875. Some five years later, the London Gazette reported in 1880 that *the Queen* [Victoria] *has been graciously pleased to appoint John Pierrepont Edwards, Esq., now British Vice-Consul at New York, to be Her Majesty's Consul at New York.* Five years later, in May 1885, at the opening of the International Inventions and Music Exhibition in London, *The Times* reported that the opening address to the Prince of

Wales [later to be King Edward VII] included the following: *Through the efforts of Mr. Pierrepont Edwards, Her Majesty's Consul at New York, a wide publicity has been given to this Exhibition in the United States, which has resulted in our having the pleasure of welcoming many exhibitors from that country.* It is thus clear that Lawrence Johnston's mother's family had close links to England.

There is little information about Gertrude before her marriage although there is an entry in the *New York Times* in May 1868 which records that the Americans registered with a firm in Paris as visitors includes Miss Gertrude Waterbury of New York. It is likely that Gertrude had developed a liking for Paris and France before her marriage two years later.

Lawrence Johnston's immediate family

Elliott Johnston and Gertrude Waterbury were married in St. Peter's Church, Westchester on 20 October 1870.

St. Peter's Church, Westchester, New York

The entry in the records of St. Peter's Church shows that Elliott was then aged 44 whilst Gertrude was 24, some 20 years younger.

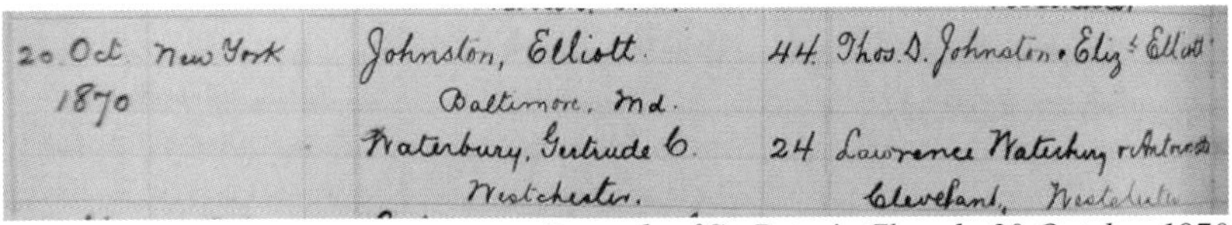
20 Oct. 1870 New York | Johnston, Elliott. Baltimore, Md. | 44 Thos. D. Johnston & Eliz. Elliott
Waterbury, Gertrude E. Westchester | 24 Lawrence Waterbury Cleveland, Westchester

Records of St. Peter's Church, 20 October 1870

A month later on 21 November, Elliott and his wife appeared before a notary to swear that he was a native citizen of the United States born in Baltimore City in order to obtain a passport. The application includes a description of Elliott Johnston.

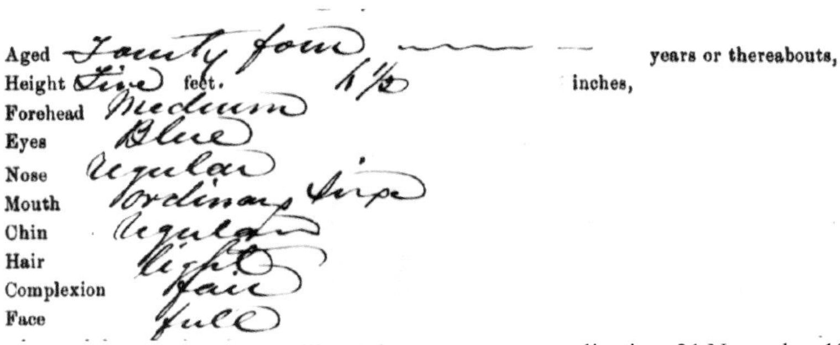
Aged Forty four years or thereabouts,
Height Five feet. 6 1/2 inches,
Forehead Medium
Eyes Blue
Nose Regular
Mouth Ordinary Size
Chin Regular
Hair Light
Complexion Fair
Face Full

Elliott Johnston passport application, 21 November 1870

A year later, Lawrence, their first child, was born on 12 October 1871 in the 16th arrondissement of Paris, France. This arrondissement is on the right bank of the Seine and is largely residential including a number of diplomatic missions and also the second largest public park in Paris, the Bois de Bologne.

The following year on 15 July 1872, Elliott, their second son, was born in the Savoy region of France. This is the region to the south of Lake Geneva and bounded on the east by Switzerland and the Mont Blanc massif and on the southeast by Italy.

And then their daughter, Elizabeth Elliott, was born on 22 December 1873 at 31, rue Pastorelli, in Nice, France. Both her parents are recorded as having the profession of '*rentier*' – or a person of independent means – and being domiciled in Baltimore, Maryland. Sadly, some seven months later in July 1874, Elizabeth died of whooping cough in Ferney, France just across the border from Geneva, Switzerland.

Elliott and Gertrude would probably spend the warm summer months in the cooler mountainous regions of France such as Savoy and the Jura close to Ferney and then return to the Mediterranean coast such as at Nice for the winter months. Elliott returned to America most years as he is included in the passenger lists of those arriving in New York in 1873, 1874 and in 1875 when the whole family travelled.

By 1880, the Johnston family were living in New York City at 3566 East 35th Street when the household consisted of:

Elliott Johnston	White Male	40 Gentleman	Married
Gertrude Johnston	White Female	30 Making House	Married
Lawrence Johnston	White Male	9 At School	Single
Elliott Johnston	White Male	7 At School	Single
Marie Franz	White Female	22 Boarder	
James Hoffmann	Black Male	25 Servant	

The Johnston family had spent several years in Europe from 1871 to 1875, with all three children being born in France and Elizabeth dying there. They may have been in the United States during the latter half of the 1870s or, alternatively, have continued living in Europe, before being in New York City in 1880.

TWO

LAWRENCE JOHNSTON'S EARLY YEARS

At the age of 9, Lawrence Johnston was living in 1880 with his father and mother and his brother in New York City on East 35th Street. The next seven years were going to see huge turmoil in his family life as his parents were to divorce and both were to remarry. In this chapter, before focussing on Lawrence's early years, I first consider the lives of his parents starting with his father's second marriage and subsequent life before considering his mother's life after her second marriage.

It is probable that Elliott Johnston was living apart from his family by 1883 if not earlier. On 5 May 1884, Lawrence's uncle, Henry, died of pneumonia at the Fifth Avenue Hotel in New York. His body was brought back to his late residence, 116 Park Avenue, Baltimore for subsequent burial at Green Mount Cemetery, Baltimore.

Henry Elliott Johnston grave (dark cross), Green Mount Cemetery, Baltimore

An article in the Baltimore *Sun* two days later noted that Henry, who had married Harriet Lane, the First Lady of President Buchanan, had been in failing health for some years. A report later that month about his will notes that bequests of $10,000 were made to his brothers and sisters – which would have included Elliott, Lawrence's father. Elliott was clearly involved in helping to sort out the affairs of his brother as in July he was reported as emphatically denying that furniture advertised to be sold in Philadelphia had ever belonged to President Buchanan but had belonged to his brother, Henry. As Elliott was described as being of Baltimore he was probably living there at that time.

The records of transatlantic crossings show that Gertrude and her sons, Lawrence and Elliott, had been in Europe as they arrived in New York from Liverpool on the *S.S. Aurania* on 5 July 1884 in an entry listing:

Mrs Elliott Johnston	Age 40	Matron	10 pieces baggage
Lawrence Johnston	Age 11	Child	
Elliott Johnston	Age 10	Child	
Constance Marn	Age 22	Maid	French nationality

They probably stayed in New York as Gertrude's mother, brother and sisters were living there.

Elliott Johnston's second marriage

An early indication of Elliott's interest in the girl that he was to marry came in 1884 when the *Peninsula Enterprise*, the local newspaper of Accomack County on the eastern shore of Virginia, on 24 May included an item recording the sale by Elliott to Alice M. Langford of *one skiff, and other personalty, at Powelton* for the sum of $200. Personalty was a word then used to describe moveable assets. The deed of sale records that these assets were:

> *one skiff, about fifteen feet long, together with its sails, oars, anchor cable and all other appurtenances; one gunning boat, together with its oars, and all other appurtenances; two breech loading guns; one muzzle loading gun; and a lot of powder, shot, shell and other*

fixtures and appurtenances used about gunning for birds or other game.

This sale reflects the vogue in the latter half of the nineteenth century in the Chesapeake Bay area for the shooting of ducks and geese, which would arrive in the annual migrations in very large numbers. The sum of $200 in 1884 was quite sizeable as $500 was enough to purchase a house and its contents and $1600 to purchase an acre and a half of land.

Powelton was a village that in the 1800s changed its name to Wachapreague. It is situated on the Atlantic coast of the eastern shore of Virginia and lies on a peninsula that is made up of Delaware and Maryland to the north and Virginia to the south. As it is situated on high ground, its access to the ocean through the barrier islands was regarded as the best on the Eastern Shore.

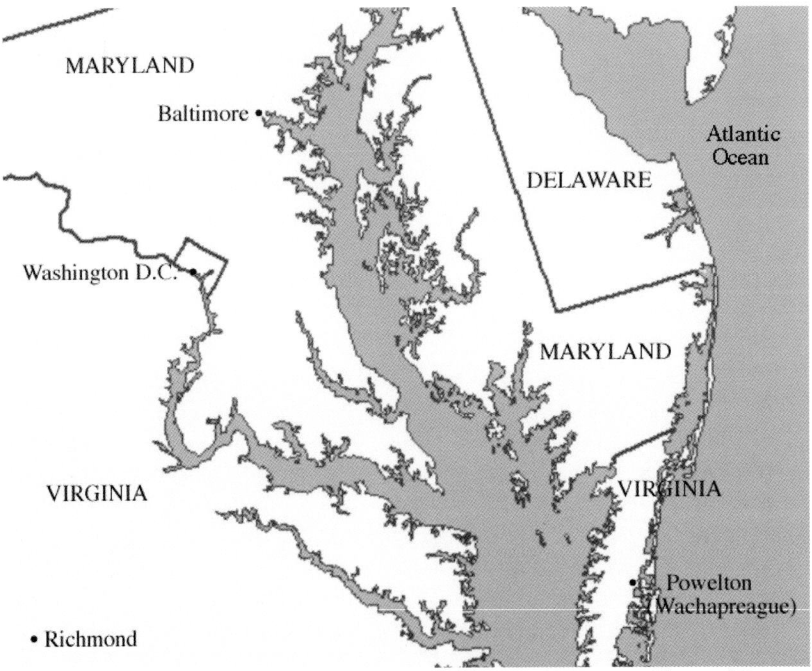

Powelton (Wachapreague) on the Eastern Shore of Virginia

Powelton was a popular village in the 1880s as the *Peninsula Enterprise* reported in September 1883 that *This new found Arcadia, it is said, will soon attract thither many other purchasers, also, and our little village by the sea assume proportions we never dreamed of.* It went on to add that *The boom in land sales continues at this place.* Another issue reported that *Thirteen lots have been sold recently for building purposes at Powelton, Va., and on most of them handsome dwellings will be erected next spring.* A first-class hotel opened at Powelton the same year and the *Peninsula Enterprise* reported that *Powelton as a sea-side resort has many attractions for lovers of gunning and fishing, and the proprietor of the Hotel, in soliciting their patronage, desires to say he will furnish them with every means to gratify their tastes in these respects.* By the 1880s, Powelton had become known as Wachapreague and had a twice-weekly freight and passenger service with New York City.

In July 1885, Elliott Johnston bought from Thomas J. Floyd a house on a plot of land of about an acre situated on the creek in *the village of Powelton or "Wachapreague City"* for the sum of $2,400 (equivalent to about $56,500 today). The deed records that:

> *... it is expressly agreed by and between the said parties that the said Elliott Johnston is to be let into possession of the said premises on the12th day of September A.D. 1885*

The reason for this is evident from the issue of the *Peninsula Enterprise* reporting the marriage on 9 September 1885 of Elliott Johnston and Alice Langford. Their marriage licence had been issued on 20 August by the Accomack County Clerk authorising *any person licensed to celebrate marriages ... to join together in the Holy State of Matrimony ... Elliott Johnston and Alice M. Langford.* The annexed certificate shows that Elliott Johnston was aged 59 and although initially shown as *Widowed*, this had been crossed out and replaced by *Divorced.* He was shown as being born in Baltimore City, Maryland and living in Accomack County, Virginia, the son of Thomas D. & Elizabeth E. Johnston, with his occupation shown as a gentleman. Alice M. Langford, age 22 and single, was also shown as being born in Baltimore City, Maryland and living in Accomack County, Virginia, the daughter of John & Maria Langford. It is likely that Alice was actually younger as her date of birth is recorded in the 1900 census as being 7 January 1867 so she would have been 18 when she married Elliott.

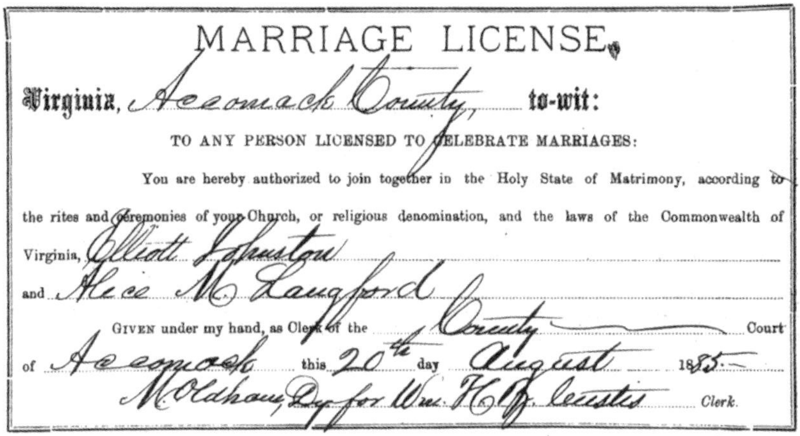

MARRIAGE LICENSE.

Virginia, Accomack County, to-wit:

TO ANY PERSON LICENSED TO CELEBRATE MARRIAGES:

You are hereby authorized to join together in the Holy State of Matrimony, according to the rites and ceremonies of your Church, or religious denomination, and the laws of the Commonwealth of Virginia, Elliott Johnston

and Alice M Langford

GIVEN under my hand, as Clerk of the County Court of Accomack this 20th day August 1885

M. Oldham, Dep. for Wm. H. B. Custis Clerk.

Marriage Licence for Elliott Johnston & Alice M. Langford, 20 August 1885

They were married by the Rev. John W. A. Elliott at Locust Mount Church, just inland from the town of Wachapreague.

Locust Mount Church, Accomack County, Virginia

The following year, Elliott and his wife entered into an interest-free mortgage with his brother, Josiah Lee, for the sum of $250 a year for the succeeding nine years totalling $2,120.

Within three years after his second marriage, Elliott was engaged in seeking to remove his *political disabilities.* These resulted from the 14th Amendment to the US Constitution – an Amendment enacted following the Civil War. It was proposed by Congress in 1866 and adopted in 1868. This reflected the determination of the northern States that the southern States should not be readmitted to the Union and the Congress without additional guarantees. The first Section made all persons born within America citizens of the United States and of the State where they resided – thus ensuring that slaves had constitutional rights. It also provided equal protection to all people. The third Section prevented the election or appointment to any Federal or State office of any person who has served as an officer of the United States and subsequently engaged in rebellion against the US:

> *Section 3. No person shall be a Senator or Representative in Congress, or elector of President and Vice President, or hold any office, civil or military, under the United States, or under any State, who, having previously taken an oath, as a member of Congress, or as an officer of the United States, or as a member of any State legislature, or as an executive or judicial officer of any State, to support the Constitution of the United States, shall have engaged in insurrection or rebellion against the same, or given aid or comfort to the enemies thereof. But Congress may by a vote of two-thirds of each House, remove such disability.*

As Elliott, when he was a Midshipman in the US Navy, had taken an oath to support the Constitution of the United States and had then subsequently served with the Confederate Forces during the Civil War, he was barred from holding any office. These political disabilities could be removed, as provided for by the 14th Amendment, by a Bill being passed by both Houses of Congress so long as a two-thirds majority of each House voted in favour. Quite often, the newspaper accounts of such bills referred to the petitioner as being an *unreconstructed rebel.* Elliott's handwritten petition to the Congress began as shown on the next page.

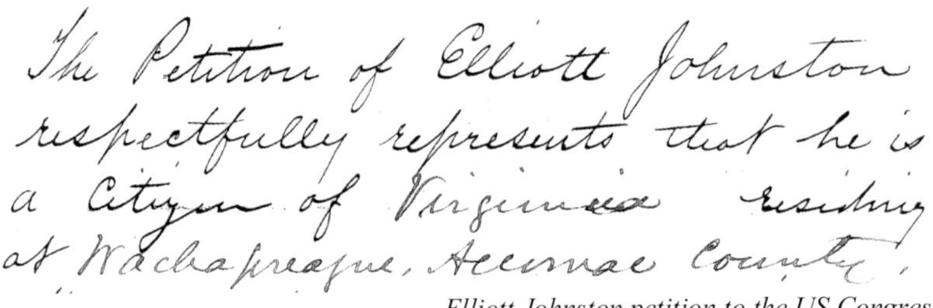

The Petition of Elliott Johnston respectfully represents that he is a Citizen of Virginia residing at Wachapreague, Accomac County.

Elliott Johnston petition to the US Congress

The petition went on to say that he had *served in the U.S. Navy as a Midshipman from Sept. 17th 1841* and that *he entered the Confederate Army in September 1861 and served therein until the close of the war.* It concludes by saying:

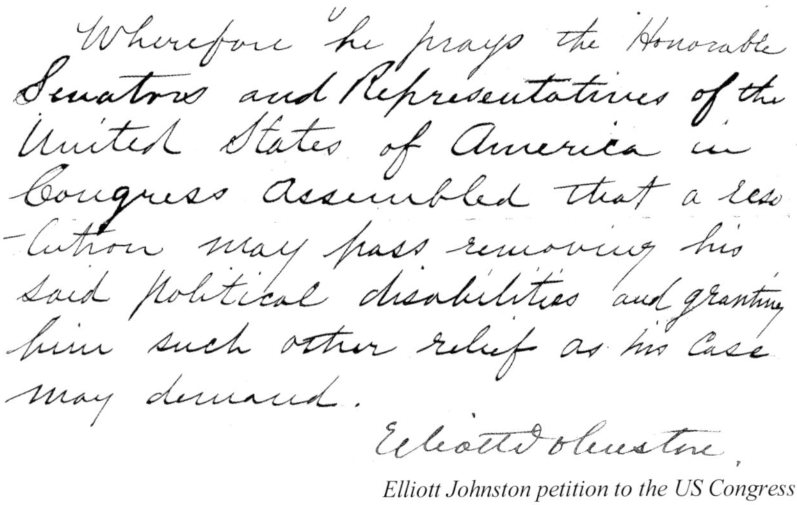

Wherefore he prays the Honorable Senators and Representatives of the United States of America in Congress Assembled that a reso-lution may pass removing his said political disabilities and granting him such other relief as his case may demand.

Elliott Johnston

Elliott Johnston petition to the US Congress

In early 1888, the Bill to remove Elliott's political disabilities was being considered by the House of Representatives as a report was published on 21 March 1888 recommending passage of this bill. The bill was considered subsequently by the Senate and in August 1888 an act to remove the political disabilities of Elliott Johnston was approved with the required two-thirds concurrence of both the House of Representatives and the Senate. Interestingly, the passage of this bill was reported as the first item in newspaper accounts of Senatorial Activity in papers such as the Philadelphia Inquirer on 1 August 1888 as follows:

> *WASHINGTON, July 31. The House bill removing the political disabilities of Elliott Johnson [sic] of Accomac County, Va., was reported and passed with a verbal amendment.* [The verbal amendment was to correct the spelling from Johnson to Johnston]

The Washington Post on 2 August 1888 carried an item entitled *Mr. Johnston's Family History* noting that:

> *Mr. Elliott Johnston, of Accomack County, Va., for the removal of whose political disabilities a bill passed the Senate Tuesday evening, is the brother of Mr. Johnston, of Baltimore, who married Miss Harriet Lane, the niece of President Buchanan, and who was the lady of the White House during the Buchanan administration.*

Gertrude's second marriage

There is little to show what Gertrude Johnston and her two sons, Lawrence and Elliott, were doing at this time. However, in 1887 on 5 November, Gertrude, then aged 41, married Charles Francis (Frank) Winthrop, aged 60, in Westminster, London at the Church of St. James located on the south side of Piccadilly. The marriage certificate describes Gertude as being *unmarried* and residing in the parish of St. James at the time of her marriage. Her husband, Frank, is described as being a *bachelor* and residing in New York, United States at the time of the wedding. The announcement of the wedding in the Baltimore *Sun* says that Gertrude had been *divorced about four years ago* – indicating a divorce in 1883 – *from her first husband, Mr. Elliott Johnston of Baltimore.* It also says that Mr. Winthrop's *marriage created considerable surprise, as it was supposed that he was a confirmed bachelor.*

Frank was born in New York on 20 March 1827 to Thomas Charles and Georgiana Maria (née Kane) Winthrop. Frank was for many years associated with his brother, Robert, in a banking and stock-broking business at No. 18 Wall Street, New York. He retired from business life some ten years before his marriage to Gertrude and at about that time took up residence in Paris, France. Interestingly, he was a direct descendent of John Winthrop, the first Governor of the Massachusetts Bay Colony. Consequently, Lawrence, already related to the First Ladies of two U. S. Presidents, was through his mother's second marriage related to yet another distinguished figure in American history.

Following her second marriage in 1887, Gertrude began to become prominent in New York society. Her name and address appeared in the *Social Register*, a directory of prominent American families that formed the social elite: *Since its inception, the Social Register has been the only reliable, and the most trusted, arbiter of Society in America.* Gertrude was included in the first edition in 1886 when she appeared as *Mrs. Elliott Johnston (Waterbury)* with no address shown. By 1891 she appeared in the following entry:

Winthrop, Mrs. Chas F (Johnston – Gertrude Waterbury) Un.	*20 W 33*
The Messrs Lawrence & Elliott Johnston	

This shows that their address in New York at that time was 20 West 33rd Street. Entries in the Social Register in subsequent years frequently show Gertrude and her sons, Lawrence and Elliott, as being abroad – as in 1898 and 1901. However, in 1903 her address is shown as being 279 Fifth Avenue, which was then the home of Mr. & Mrs. Buchanan Winthrop, with Lawrence being abroad and Elliott in the US. In summer 1905, Gertrude is shown as being at *The Antlers*, Raquette Lake, New York – a hotel in the Adirondacks in upstate New York, one of the places to which New York society went in the summer months. A year later, in summer 1906, Gertrude's address is the *Villa Sorrento* at Dinard, France. Then in November 1907, although shown as being abroad her address is given as 40 East 69th Street, New York City – a property which Gertrude continued to own up until her death in 1926.

As well as appearing in the Social Register, Gertude's activities were also recorded in the social column of the *New York Times*. An early entry under *Society Events of the Week* was on 1 February 1891 which noted that the large hotels in St. Augustine, Florida have opened their doors and that *The Florida season promises to be an unusually gay one* with the recent arrivals including Mr. and Mrs Frank Winthrop and Mrs Lawrence Waterbury (Gertrude's mother). Another entry in December 1891 notes the Charity Ball to be held at the Metropolitan Opera House in

New York saying – *as an opportunity to see the notable men and women of this city is unrivalled* – and including Gertrude among the numerous patrons. An entry for 25 March 1892 lists those who attended Gertrude's reception at her home at 20 West Thirty-third Street on Wednesday afternoon when Mme. Pemberton-Hincks, Ross David and William Lindberg sang. Mme. Pemberton-Hincks was a soprano who sang in the opera *Cavalleria Rusticana* performed under the direction of Oscar Hammerstein. Two days later, a list is given of those who attended as guests of Gertrude on Thursday afternoon.

These entries continue with one on 4 May 1892 advising that *An afternoon tea will be given by Mrs. Charles F. Winthrop of 20 West Thirty-third Street today. A large number of cards have been issued.* Two days later, Gertrude is listed as one of the attendees at the Claremont tea held on 5 May which noted that *At 4:30 o'clock, the tea hour, the broad piazzas of the café comfortably seated 100 of the 150 subscribers* and went on to add that *While tea was served, Lander's Orchestra played.* Six months later, on 27 November 1892 the opening ball at Lakewood, New Jersey included Gertrude – an occasion described as follows: *Wealth, beauty and fashion combined to render the opening ball at the Lakewood tonight one of the most brilliant events in the history of the hotel and of the community.* Interestingly, at another ball some three weeks later on 13 December 1892 at Delmonico's in New York City, both Gertrude and Lawrence were present at an occasion at which *the Patriarchs were the hosts of the wealth and beauty of New York.* It was noted that *The 230 persons were seated at thirty tables. The ballroom was decorated with palms and in the centre of each table was a bouquet of specimen American Beauty roses. During the dinner Lander's Orchestra played in the balcony.*

Two months later, on 26 February 1893, Gertrude was a recent arrival from New York at St. Augustine, Florida. This is a city on the Atlantic Coast of Florida which was a popular winter resort for the wealthy in the late nineteenth century as it enabled them to escape from the cold winter temperatures of New York. A few days later on 5 March 1893 Gertrude was a guest at *a larger dinner at the Villa Zorayda one evening last week.* The society engagements continue with another example being on 9 August 1895 when Gertrude attended a ball for a debutante at Bar Harbor, Maine. Bar Harbor was a popular place for New York society to spend the summer months as it is on the Atlantic

Coast offering relief from the summer heat of New York City. By 1880 there were some 30 hotels and wealthy visitors would rent cottages – which were large houses with extensive accommodation for residents and their staff.

Bar Harbor, Maine in the late 1800s

The pace of Gertrude's social activities in her fiftieth year was strenuous as her attendance at further balls at Bar Harbor was reported on 14 and 18 August and again, on 21 August 1895. Six days later, on 27 August 1895 there are no less than three mentions of functions attended by Gertrude – the first again a dinner and dance whilst the second noted that she had attended *a delightful musicale at Cloverfoot this morning, which was attended by a large delegation of society folk* and the third was that she had been to a lecture on *Romans de Chopin* attended by *a number of prominent society people*. The following day, 28 August 1895, noted that:

> *Mrs Charles Francis Winthrop of New York gave a luncheon party at Kebo today, which for beauty of floral decorations has seldom been eclipsed. A large round centerpiece of hardy hydrangea and gladioli was tastefully arranged inside a bank of wild ferns and smilax.*

Corsage bouquets of American Beauty roses tied with wide ribbon of cream-coloured silk were given to the ladies present.

There was no let up in the social whirl as the next day, 29 August 1895, noted Gertrude had been to the Wednesday dinner and dance at Kebo Valley Park and that she had also attended an art conference. Two days later, she attended a tea reception and then the following week Gertrude gave a dinner party at a Bar Harbor hotel at which the guests included Baron Hengelmuller von Hergervar, the Austrian minister and his wife, and four days later she attended an afternoon tea. It is probable that her actual social engagements were much more numerous as it is unlikely that all would be reported in the Social Column of the *New York Times*.

Society in New York in the late 1800s typically went north to Bar Harbor, Maine for the summer months when it was too hot in New York City, then in the late autumn went to Lakewood, New Jersey and immediately after Christmas went south to St. Augustine, Florida returning to New York in the spring.

It is now time to consider what had been happening to Lawrence Johnston since his mother's marriage to Charles Francis Winthrop in London on 5 November 1887. An entry in the Alumni Directory 1932 for Columbia University – which is located in New York City – shows an entry for Lawrence:

Johnston, Lawrence *x*1894 Arch,
+

and also one for Elliott:

Johnston, Elliott jr *x*1895 E, +

The symbol *x* indicates that neither graduated; the 1894 and 1895 indicate that they were in the classes of 1894 and 1895; the Arch indicates that Lawrence enlisted in the School of Architecture whilst Elliott, Jr. enlisted in the School of Engineering earlier known as Applied Science and the Bureau of Mines, Engineering and Chemistry; and the + indicates address unknown, a symbol frequently used for alumni of earlier years for which no alumni records were compiled. Although the Columbia University records have no information on the addresses for these Johnstons, it is probable that they are the children of Elliott and Gertrude as they were born in 1871 and 1872 respectively and would have been expected to go to university in successive years. Gertrude's brother, James Montevaudert Waterbury, had graduated with an A.B. from Columbia in 1873 and could well have encouraged Gertrude to send her sons to Columbia. Elliott Johnston was in Berlin in October 1892 as he applied for a passport for the purpose of *studying in Berlin.* His application says that his permanent residence is in New York where he is a student and that he had left the United States on 21 September 1892 and intended to return to the United States within two years.

On 6 July 1893, Gertrude and Lawrence landed at Southampton on the *S.S. Fürst Bismarck* from New York. This photograph showing Lawrence and his mother will have been taken at about that time.

Lawrence Johnston and his mother, Gertrude Winthrop

A month after arriving in England, in August 1893, Lawrence enrolled in a small private school run by John Dunn, M.A. at Kirby Lodge, 32 High Street in Little Shelford, a village five miles south of Cambridge.

Kirby Lodge, Little Shelford, Cambridgeshire

There were usually four students enrolled at a time who would be tutored by John Dunn and his son, Arthur, to enable them to reach the standard required for entry into colleges in the University of Cambridge. Whilst attending Kirby Lodge from August 1893 to October 1894, Lawrence enrolled at Trinity College, Cambridge as a pensioner on 25 June 1894.

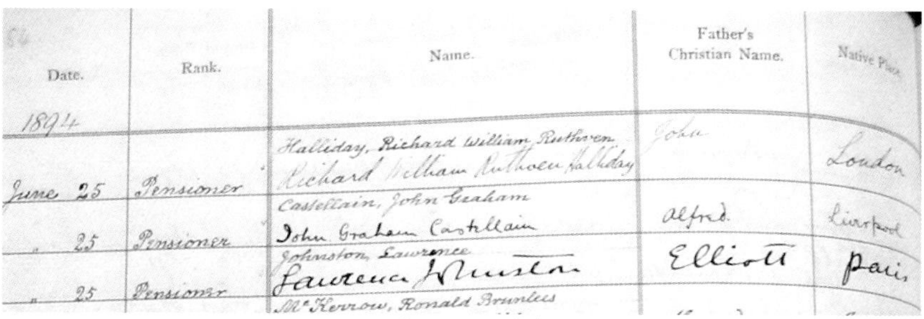

Date.	Rank.	Name.	Father's Christian Name.	Native Place
1894				
June 25	Pensioner	Halliday, Richard William Ruthven Richard William Ruthven Halliday	John	London
" 25	Pensioner	Castellain, John Graham John Graham Castellain	Alfred	Liverpool
" 25	Pensioner	Johnston, Lawrence Lawrence Johnston	Elliott	Paris
		McKerrow, Ronald Brunlees		

Lawrence Johnston entry, Trinity College records

Lawrence was at Trinity College for three years from October 1894 to October 1897 when his tutor was Walter William Rouse Ball, a mathematician who was a fellow of the College from 1878 to 1905.

Trinity College, Cambridge

Lawrence's record card shows that he matriculated on 22 October 1894 and then went on to take the Part II of his Previous exams in the October Term 1894 followed by Part I in the Easter Term 1895. He then took the Part II of his General Examinations in the Michaelmas Term 1895 with Part I in the Easter Term 1896. Finally, he sat his Special examinations taking History II in the Easter Term 1897 and History I in Michaelmas Term 1897.

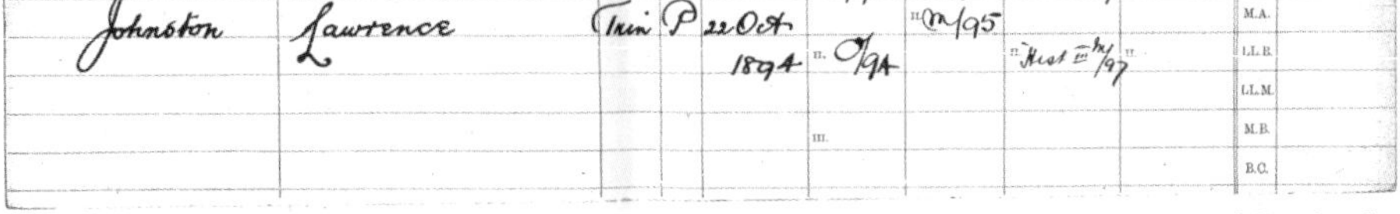

Surname	Christian Name	College	Rank in Coll.	Matriculation	Previous	General	Special	Tripos	Degrees
Johnston	Lawrence	Trin	P	22 Oct 1894	I. E/95	I. E/96	I. Hist II E/97	I.	B.A.
						II. M/95			M.A.
					II. O/94		II. Hist III M/97	II.	LL.B.
									LL.M.
					III.				M.B.
									B.C.

Lawrence Johnston, Record Card, University of Cambridge

He was thus complying with the requirements set out in the *University Calendar* for students wishing to obtain an Ordinary B. A. degree. They have to sit a Previous Examination held three times a year, then a General Examination held twice a year and finally a Special Examination in one of a number of subjects – of which History was one. It is stated in the Calendar that all of these examinations are taken in two parts that can be sat either together or one part at a time.

The examination papers sat by Lawrence are available in the University Library at Cambridge which also has records of his passes in these various examinations:

Previous Part II October 1894	Fourth Class
Previous Part I June 1895	Fourth Class
General Part II December 1895	Class III
General Part I June 1896	Class IV
Special (History) Part II June 1897	Class II
Special (History) Part I December 1897	Class III

The papers that he sat for the Special Examination in History in 1897 included one on the *History of the Papacy*.

WEDNESDAY, *December* 1, 1897. 1—4.

HISTORY OF THE PAPACY.

1. What do you know of the career of Charles of Durazzo?

2. The pontificate of Urban VI "is one of the most disastrous in the whole history of the Papacy."

Give evidence in support of this statement.

3. What part did the University of Paris take in the attempts to restore the unity of Christendom?

4. Describe the steps by which Ladislas finally succeeded in making himself master of Rome.

Special Examination in History, University of Cambridge, December 1897

The Times on 10 December 1897 reported that on the previous day the degree of Bachelor of Arts had been conferred on Lawrence Johnston of Trinity College showing that he had successfully completed his three year degree course at Cambridge. Throughout his time at Trinity College, Lawrence lived in lodgings at 12, Portugal Street, Cambridge where Mrs. Eliza Cherry was the lodging keeper.

12, Portugal Street, Cambridge

Following his graduation, Lawrence went north in 1898 to New Etal in Northumberland to become a farming pupil.

Death of Lawrence's step-father

His step-father, Charles Francis Winthrop, died in Paris, France on 16 February 1898. The Pall Mall Gazette on Saturday, 19 February, reported that *A funeral service will be held today in the American Church in the Avenue de l'Alma over the remains of Mr. Charles Francis Winthrop, one of the oldest members of the American colony in Paris. The body will be conveyed to New York for interment.* His death certificate in the Archives of Paris show that he died at Rue Castiglione, 3 which is today one of the top hotels in Paris, the Westin Hotel, and was the Hotel Continental in the 1890s. It had been built in 1878 and had quickly become the preferred residence of countless royals and celebrities. It overlooks the Tuileries Gardens and with the Eiffel Tower beyond. It is likely that Frank Winthrop had one of the 78 private apartments in the Hotel.

Westin Hotel, previously the Hotel Continental, Paris

His obituary in the *New York Times* on 17 February 1898 recorded that he was born in New York in 1827 and *was descended from one of the oldest families in New York. When a young man he became connected with the Wall Street firm of Drexel, Winthrop & Co.* and remained with them for many years. Drexel, Winthrop & Co. were bankers and stockbrokers with a thriving business in New York and Philadelphia during the 1860s when they had an annual profit of around $300,000 a year (equivalent to about $4 million a year today). His obituary notes that *twenty years or more ago he retired from business life, and ten years later he took up residence in Paris.* The obituary went on to note that *His widow survives him.*

Frank's will, which was made on 9 June 1893, said that he was of New York City, *now temporarily residing in Paris.* He left all his estate to his sister, Gertrude Winthrop, for her life and then, after her death, to his nieces. Nothing was left to his wife, Gertrude, or to his step-sons, Lawrence and Elliott, Jr.

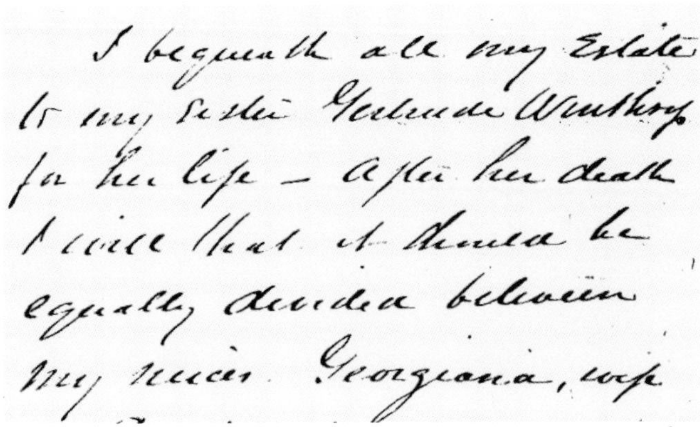
I bequeath all my Estate
to my sister Gertrude Winthrop
for her life – after her death
I will that it should be
equally divided between
my niece Georgiana, wife

Extract from the will of Charles Francis Winthrop

Death of Lawrence's father

Lawrence's own father, Elliott, was also not well and was approaching the end of his life. A report in the Baltimore *Sun* on 7 May 1900 said that:

> *Major Elliott Johnston, formerly of Baltimore, but whose home is now on the Eastern Shore of Virginia, was operated upon several days ago for the removal of a tumor at the Maryland University Hospital, Lombard and Greene Streets. ... Major Johnston is getting along very nicely. Within several days he expects to leave the hospital and spend a short time with his brother, Mr. Josiah L. Johnston, 113 West Franklin Street, before returning to Virginia. Major Johnston was aid to Lieut. Gen. Ewell in the Civil War. Many of his old friends and relatives have called to see him since he has been in Baltimore.*

Elliott was back in Wachapreague by 4 June 1900 when the US census showed that the household consisted of:

Johnston, Elliott,	*head,*	*born 1 May 1826,*	*aged 74*
Alice M,	*wife,*	*born 7 January 1867,*	*aged 33*
Hood, May,	*servant,*	*born 5 May 1882,*	*aged 18*

However, he was probably not well after his operation – he was by then 74 – and three months later, on 23 August, he sold his house and land and all his belongings in Wachapreague, Accomack County, to his wife Alice M. Johnston for the sum of one dollar. The deed states:

> *This deed made this 23rd day of August, A.D. 1900, between Elliott Johnston, of the one part, and Alice M. Johnston, of the other part, both of Accomack County, in the State of Virginia, <u>Witnesseth</u>: that the said Elliott Johnston, for and in consideration of the natural love and affection which he has and bears for the said Alice M. Johnston, and for the further consideration of the sum of one dollar ...*

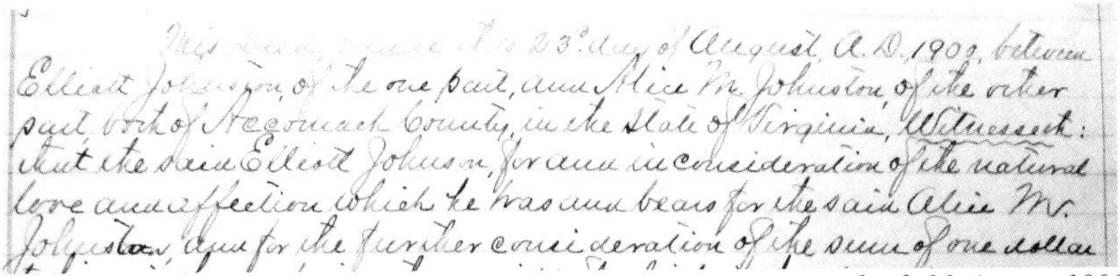
This deed made this 23rd day of August, A.D. 1900, between
Elliott Johnston, of the one part, and Alice M. Johnston of the other
part both of Accomack County, in the State of Virginia, Witnesseth:
that the said Elliott Johnston, for and in consideration of the natural
love and affection which he has and bears for the said Alice M.
Johnston, and for the further consideration of the sum of one dollar

Elliott Johnston deed, 23 August 1900

The deed also includes *all that certain lot, piece or parcel of land containing by estimation one acre (1 a), more or less, situate and being in the village of Wachapreague, in the said County, ... together with all the improvements, privileges and appurtenances to the said land belonging or appertaining;* It then details the animals and carriages and furniture that were also sold to his wife Alice, thus providing an interesting insight into their way of life at Wachapreague:

> *one sorrel mare called "Crickett"; one bay mare called "Lady Alice"; one brown mare called "Thelma"; one cow and calf; one top buggy; one speed cart; one double-seated road cart; one four-seated carriage; one buck board; one lot of harness; all the household and kitchen furniture of the said Elliott Johnston, of whatsoever kind and description, and one piano;*

Some five months later, on 28 January 1901, the Baltimore *Sun* reported that *Major Elliott Johnston, whose severe illness was noted in The Sun a few days ago, is not in an improved condition, and his advanced age, 75, renders recovery out of the question.* The article goes on to summarise Elliott's involvement in the Civil War by saying:

Major Johnston was one of the many gallant Maryland men who hurried to the aid of the South in 1861. He served on the staff of Angus McDonals, of Richard B. Garnett and of R. S. Ewell and was severely wounded at the battle of Sharpsburg, losing his left leg. He returned to duty on recovery and went through the Gettysburg campaign with General Ewell as his assistant adjutant-general. He remained in active service until incapacitated for further field duty by renewed trouble with his wound and the prospect of a second amputation.

The following day the Baltimore *Sun* reported that *Major Elliott Johnston, a gallant soldier of the Confederacy, died yesterday morning at St. Joseph's Hospital, Caroline and Hoffman streets, where he had been under treatment since last October.* The obituary goes on to say that he was *the son of Thomas Donaldson Johnston, a well-known banker of this city.* and in a resume of his life records that *At the age of 14 he entered the United States Navy as a midshipman, resigning several years later. He was of a roving disposition and prior to the Civil War he visited many foreign countries.* It concludes by noting that he is *survived by two sisters, Mrs. Thomas Baxter Gresham and Miss Mary W. Johnston, 815 Park Avenue, and a brother, Mr. Josiah Lee Johnston, 113 West Franklin Street.*

Three days later, on 4 February, the Baltimore *Sun* reported that the funeral had taken place on Saturday morning when Major Johnston was interred in the Johnston family vault in Green Mount Cemetery with the Episcopal Church service being read by three ministers, the Rev. F. H. Stubbs and Rev. J. Woods Elliott of St. John's Protestant Episcopal Church, Waverly, Baltimore and the Rev. Robert H. Paine, of Mount Calvary Church, Baltimore. The *Peninsula Enterprise* in its report of Elliott's death said that *He had many friends on the Eastern Shore who will learn with sorrow of his death. He was a gallant Confederate soldier.* The note went on to reprint what had been said in the Baltimore *Sun* and included an extract from a letter General Garnett had written to Elliott's family when he was wounded at Sharpsburg saying *In the several battles of Kernstown, Boonsboro and Sharpsburg he was conspicuous for his gallantry and fearless bearing. I am acquainted with no-one who bears a higher or more enviable reputation than himself for courage and daring among all those with whom he has served.*

Johnston family tomb, Green Mount Cemetery, Baltimore

His will, written on 23 August 1900, the same day as the deed when he sold his home and land and all his furniture to his wife, Alice M. Johnston, for the sum of one dollar, left *to my present beloved wife Alice M. Johnston and her heirs for ever all my estate, real, personal and mixed, of every kind and description, and wheresoever situate and being.* Nothing was left to either of his two sons, Lawrence and Elliott Jr., by his first marriage to Gertrude Waterbury.

Very soon after Elliott's death, his widow, Alice, put a note in the *Peninsula Advertiser* to say that on 16 February 1901 she would *sell at public auction, at the home in Wachapreague, valuable personal property, consisting in part of household and kitchen furniture, horses, carriages, etc.* Then three months later, in April 1901, Alice sold the home at Wachapreague and the land on which it stood for the sum of $2,800 (equivalent to $70,000 today) to Charles Boulter of Philadelphia. At the same time, Alice advertised in the *Peninsula Enterprise* that she offered *for private sale two first-class double poles with yokes, also one buckboard wheelbarrow, 1,500 shingles, bricks, lumber etc.* She put her name as Mrs. Alice M. Johnston, Baltimore, Maryland suggesting that she had returned to Baltimore after Elliott's death.

It is likely that the reason why Elliott sold all his land and belongings to his second wife, Alice, for $1 on 23 August 1900, the same day as he made his will leaving everything to her, and with one of the witnesses being the same, was to pre-empt any attempt by Gertrude to challenge his will, on either her behalf or that of their sons, as the estate transferred under the will would have amounted to very little.

Lawrence Johnston in Northumberland

Lawrence went north in 1898 to become a farming pupil with George Laing, a son of Sir James Laing, at The Grange, New Etal, near Cornhill on Tweed, Northumberland. The census information for 1901 shows that those living at The Grange then were:

George Laing,	Farmer	aged 34
Annie M.	Wife	aged 37

together with a household consisting of a cook, a housemaid, parlourmaid, kitchenmaid and coachman. George Laing's father, Sir James, was a shipbuilder, who had been a director of the Suez Canal Company, whose Northumberland residence was Etal Manor. The Grange is a substantial farmhouse.

The Grange, New Etal, Northumberland

It would have been regarded as a model farm in the 1890s.

The Grange, New Etal, Northumberland

In January 1900, Lawrence applied to become a naturalized British citizen. His formal petition to do so was completed and sworn in front of a Justice of the Peace in Lincoln's Inn, London on 10 January 1900. In this he declares that he intends to reside permanently within the United Kingdom and that he wishes to become a British citizen from *his desire to serve in the Imperial Yeomanry about to be sent to South Africa.* The petition requires that the applicant shall have lived in the United Kingdom for at least five of the last eight years. Lawrence declares that he was at Little Shelford from August 1893 to October 1894, then at Trinity College, Cambridge from October 1894 to December 1897, and then again in Little Shelford from December 1897 to October 1898 before going to live at New Etal, Northumberland from October 1898 onwards.

Five referees, all British citizens and householders, were required to support the petition and had to swear that they knew Lawrence Johnston and for how long they had known him. They were:

George Laing of New Etal – 3 years 6 months
Walter William Rouse Ball of Trinity College – more than 5 years
George Frederick Oakden Bagnall of Little Shelford – over 6 years
Vandeleur Bright Bright Smith of Little Shelford – 6 years
Alexander Dick-Cunyngham of 15 Ecclestone Square, London – over 8 years

George Laing was the owner of the farm at New Etal where Lawrence was a farming pupil and he had known Lawrence since the summer of 1896. Walter Ball was Lawrence's tutor at Trinity College who certified that he had known Lawrence from October 1894 to December 1897 and had also seen him at intervals during the period from July 1893 to October 1894 when Lawrence was at Little Shelford. George Bagnall and Vandeleur Bright were both friends of Lawrence from the time that he lived at Little Shelford whilst Alexander Dick-Cunyingham had been a friend for over eight years.

It is interesting to see how quickly Lawrence's application was dealt with. He had completed his petition on Wednesday 10 January and then Laing swore his declaration in Newcastle-upon-Tyne the following day, Ball, Bagnall and Smith in Cambridge on Saturday 13 January, and Dick-Cunyngham in London on Tuesday 16 January. The Home Office received the application the same day and sent it off on the following day to the Chief Constable of Northumberland to inquire into the applicant and his referees. The reply on Saturday 20 January states that a superintendent had visited New Etal and found that Lawrence Johnston and George Laing were from home as both were at Newcastle-on-Tyne *undergoing a course of training at the Yeomanry School there.* It also reported that George Laing's wife had said that *Mr Johnston has resided with her husband since October 1898 and that he is a gentleman of great respectability and bears an excellent character.* The formal application was then forwarded to the Chief Constable of Newcastle-upon-Tyne who on Tuesday 23 January advised the Home Office that *Mr. Johnston has been seen and he says that he intends to reside permanently in the United Kingdom and that his reason for applying for a Certificate of Naturalization is his desire to serve with the Imperial Yeomanry in South Africa.* Two days later on Thursday 25 January, just fifteen days after Lawrence Johnston signed his application, the Home Office issued his Certificate of Naturalization and this was reported in the London Gazette.

THE NATURALIZATION ACT, 1870.

LIST of ALIENS to whom Certificates of Naturalization or of Readmission to British Nationality have been granted by the Secretary of State under the provisions of the Act 33 Vic., cap. 14, and have been registered in the Home Office pursuant to the Act during the Month of January, 1900.

Name.	Country.	Date of Certificate.	Place of Residence.
Johnston, Lawrence ...	United States of America	25th January, 1900 ...	Northumberland, New Etal, Cornhill-on-Tweed

London Gazette, 2 February 1900

Lawrence's military record shows that on 11 January 1900, the day after he had made his declaration in London for naturalization, he enlisted in the Imperial Yeomanry as a private (3,296) as a member of the 15th (Northumberland and Durham) Company. His enlistment form shows that he has been a *farm student* with G. Laing, Esq., at New Etal for 15 months. The medical record forming part of Lawrence's papers shows that he was aged 28, 5 ft 7¾ in tall, weight 9 st. 6 lbs and a Roman Catholic.

Lawrence sailed for South Africa as part of the 5th Regiment of the Imperial Yeomanry on the *S.S. Monteagle* on 2 February 1900. The 14th and 15th Companies returned to England after trekking for some 2,900 miles across South Africa in June 1901. Lawrence was promoted to Second Lieutenant on 12 October 1901 and a week later given temporary promotion to Lieutenant. He returned to South Africa to serve in the 26th Battalion of the Imperial Yeomanry in November 1901. He arrived back from South Africa on the *Braemar Castle* arriving on 28 August 1902. A week after his return, he was reappointed Second Lieutenant in the Northumberland Hussars Yeomanry and at the end of the month was appointed an Honorary Lieutenant in the Army.

The Northumberland Hussars were a territorial force with its members serving in the wars and in between attending summer camps lasting from eight to sixteen days. This Annual Training Camp was held at Rothbury, Northumberland from 1900 to 1904 and then on several occasions at Walwick Grange, near Fourstones, Hexham and later at other places in Northumberland and Yorkshire.

Lawrence Johnston sitting on steps on left, Northumberland Hussars annual camp, ~ 1903.

Whilst Lawrence was serving with the Imperial Yeomanry in South Africa, his mother was in Berlin as the *New York Times* on 11 March 1900 in an item headed *Americans in Berlin* mentions that Gertrude Winthrop will give a musicale on Thursday. Another entry in the same item says that *Elliott Johnson* [sic] *has passed his examination as a naval architect.* This is probably a reference to Lawrence's brother, Elliott, who had been studying in Berlin in 1892, as this item's appearance in the column *Americans in Berlin* just below Gertrude's musicale could result if she had provided both pieces of information to the *New York Times*.

Gertrude's social engagements continued apace with the *New York Times* including several references during 1901 and 1902: a Saturday night dance at Tuxedo, New York State on 2 November 1901 and the last musical morning of the season in New York on 28 June 1902. Another item on the same day, says that Gertrude has taken a residence, Sunnyside, at Bar Harbor, Maine for the season and will give a dinner there on Saturday evening. This mentions that Gertrude has not been at Bar Harbor for many years. The following day, the *New York Times* says that the marriage of Miss Julia Pierrepont Edwards, daughter of Gertrude Winthrop's sister, Antoinette Edwards, and Reginald Mansfield Johnson of Boston will be the leading social event of Bar Harbor next week. The ushers for the wedding will include Elliott Johnston of New York – the brother of Lawrence. Gertrude attended the wedding on 3 July and then the following day she gave a dinner followed by dancing in honour of her niece, Miss Julia Pierrepont Edwards, now Mrs Reginald Mansfield Johnson and the entire bridal party. Then in November, Gertrude was reported as having been staying at Chatsworth, New Jersey and then on 2 December 1902 she attended the first musical morning in the Waldorf Hotel, New York. Two weeks later, on 17 December, it was reported that Gertrude of 46 West Forty-seventh Street, New York had given a luncheon. Then after Christmas, the *New York Times* reported that she was staying at the Lakewood Hotel, Lakewood, New Jersey.

Lawrence, as already noted, returned from South Africa in August 1902. He took part in the Northumberland Hussars annual camps at Rothbury on 20 July/3 August 1903 and again on 24 June/8July 1904. He probably returned from the Boer War to Northumberland but then a year or so later, he moved back to Little Shelford, Cambridgeshire.

Although Lawrence had been a faming pupil at New Etal, it is evident that he was developing an interest in gardens in the early 1900s as the Journal of the Royal Horticultural Society records that at a General Meeting on 5 April 1904, he was one of the 40 Fellows elected on that occasion.

GENERAL MEETING.

April 5, 1904.

Mr. Bunyard, V.M.H., in the Chair.

Fellows elected (40).—C. Allen, G. F. O. Bagnall, Miss Gertrude Boulter, Alfred Brisco, Mrs. H. de V. Brougham, Mrs. A. H. Burton, John Butler, D. A. Christie, Mrs. C. F. Churchill, Lady Cowell, Andrew Crane, Howard Dickinson, Eric Drabble, W. Emerton, Hon. K. Forbes-Sempill, Mrs. H. F. Fox, A. Freeland, Ruben Frogbrook, S. W. Fryett, F. A. Hooper, Lawrence Johnston, Sir William Johnston, Bart., Mrs. H. Lamotte, Colonel H. C. Legh, Mrs. Norman Leslie, A. H. Louis, Miss I. Maudsley, Mrs. A. Meysey-Thompson, Lady Nicholson, B. Othmer

Journal of the Royal Horticultural Society, Vol. XXIX, xix, 1905

Interestingly, another Fellow elected at the same time as Lawrence, was G. F. O. Bagnall, who was one of Lawrence's friends from his first stay at Little Shelford in 1893, and had sworn a declaration in support of

Lawrence's naturalization in 1900. Bagnall, however, was only a Fellow of the R.H.S. for a year. Lawrence is in the List of Fellows for 1905 as:

Johnston, Lawrence, Little Shelford, Cambs.

He then started to borrow books from the R.H.S. Lindley Library initially in the summer of 1905 on alpines:

Alpine Plants, W. A. Clark
Les Plantes des Alpes, H. Correvon
Alpine Flora, Hoffmann
Alpine Flowers for Gardens, Robinson
Atlas de la Flora Alpines (5 Tomes & Index)

and then in autumn 1905 on plant breeding and on gardening in general, including *The Art and Craft of Garden Making* by Thomas H. Mawson:

Plant Breeding, L.H. Bailey
Book of the Wild Garden, S. Fitzherbert
Book of Climbing Plants, S. Arnott
The Book of Bulbs, S. Arnott
History of gardening in England, Amherst
Garden Craft, Mawson
Century Book of Gardening, E. T. Cook

These loans throughout 1905 show that Lawrence was developing a keen interest in gardening.

In addition, in May 1905, after qualifying at the School of Musketry at Hythe, Kent, he was promoted to Lieutenant in the Northumberland Hussars.

IMPERIAL YEOMANRY.

Northumberland (Hussars); **Captain D. L. Selby-Bigge to be Major. Dated 27th May, 1905.**

Second Lieutenant L. Johnston to be Lieutenant. Dated 27th May, 1905.

Lieutenant J. R. I. Hopkins resigns his Commission. Dated 27th May, 1905.

London Gazette, 30 May 1905

and then he attended the annual camp at Walwick Grange from 3 to 16 June 1905.

During the early 1900s, Gertrude was continuing her social life. She sailed to New York in June 1904 arriving on 22 June as the *New York Times* reported that she had arrived on the *Kaiser Wilhelm*. She then went to Bar Harbor, Maine when it was reported on 9 August that Miss de Garmendia is at Bar Harbor, Maine with Gertrude for the remainder of the summer – Miss de Garmendia was the daughter of an old New York family which had close links to the Queen of Spain. Some ten days later, Gertrude was among those who entertained at dinner at the weekly dinner dance at the Malvern, Bar Harbor. She had entertained a party of ten, in honour of Miss Gladys Giddings of Baltimore, which included several officers from the French cruiser *Duplex*. The following week, on 26 August 1904, she hosted a dinner for 18 at the Malvern, Bar Harbor, before the weekly dance. Early in September, Gertrude hosted a luncheon at the Malvern on Saturday at noon and on Thursday evening a dinner was given for her.

Malvern Hotel, Bar Harbor, Maine

Lawrence, together with his valet, Ernest Canham, arrived in New York on the S. S. *Baltic* on 28 January 1905. The record for his arrival shows that his address in England was Little Shelford and that his previous visit to the United States had been in 1903. The entry for his valet records that his ticket had been paid by Lawrence and that he had not previously visited America. Lawrence was back in England by April as his loans from the RHS Library then resumed.

The *New York Times* shows that Gertrude's social life was continuing in New York City as on 18 February 1905 she attended a musicale and on 5 March she was a box holder at an evening concert given by Mr. Bagby in the ballroom of the Waldorf-Astoria. Two further mentions of her and the Bagby concert appear on 9 and 10 March 1905. Later that year, on 10 September 1905, the *New York Times* reported that Gertrude had entertained at the Muenchinger-King cottage at Newport, Rhode Island.

Lawrence, again with his valet, whose name is now recorded as Edward David Canham, sailed from Liverpool on the S. S. *Baltic* on 29 November and arrived in New York on 8 December 1905 to stay at his mother's home at 40 East 69th Street. The sailing date again fits with the loans from the RHS Library as the last ones borrowed in 1905 were returned on 11 November.

The next social engagement reported for Gertrude is early in January 1906 when it was noted that she had sent out invitations for a large luncheon and card party on 20 January 1906.

Some six months later, on 4 July 1906, the *New York Times* reports that Gertrude has sailed for Europe and will be returning in the autumn to take possession of her new residence in East Sixty-ninth Street, New York.

Meanwhile, Lawrence was clearly busy with the Northumberland Hussars as *The Times* reported in its Court Circular that on 10 July 1906, King Edward VII and Queen Alexandra had made a visit to Alnwick Castle, Northumberland and had opened a new high-level rail bridge over the Tyne. It also reported that a Royal Procession had been formed and was accompanied by an escort of the Northumberland Hussars under the command of Lieutenant L. Johnston.

A Royal Procession was formed and proceeded towards Alnwick Castle, escorted by an Escort of Northumberland Hussars under the command of Lieut. L. Johnston.

The Times, 10 July 1906, p. 10

The Regimental Orders of the Northumberland Hussars show that Lawrence was in charge of the Royal Escort from Sunday 8 to Thursday 13 July 1906.

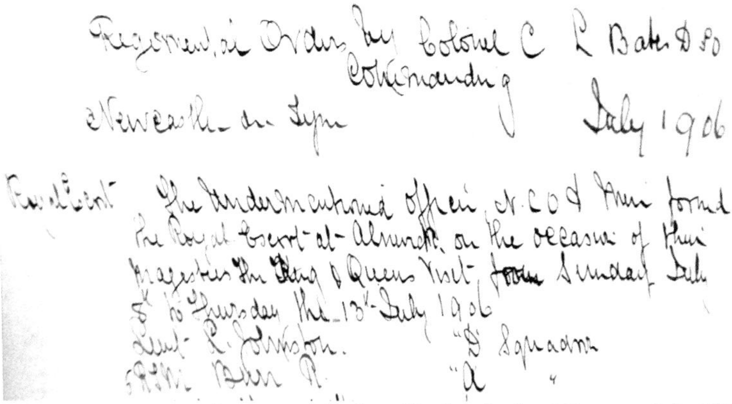
Regimental Orders by Colonel C L Bates D.S.O. Commanding
Newcastle on Tyne July 1906
Royal Escort The undermentioned officer, N.C.O. & men formed the Royal Escort at Alnwick on the occasion of their Majesties the King & Queen's visit from Sunday July 8th to Thursday the 13th July 1906
Lieut. L. Johnston. "D" Squadron
S.S.M. Brown R. "A" "

Regimental Orders, Northumberland Hussars, July 1906

Lawrence is on the white horse leading the escort:

Royal Escort, Lieut. L. Johnston in charge, July 1906

In April of the following year, 1907, Lawrence, some 5 months before coming to Hidcote, borrowed three further books from the Lindley Library. These included two of Gertrude Jekyll's books *Home and Garden* and *Wood and Garden* as well as A.B. Freeman-Mitford's book *The Bamboo Garden.* Interestingly, Algernon Bertram Freeman-Mitford was the first Baron Redesdale who lived at Batsford, where he created much of the Arboretum, some 10 miles away from Hidcote.

The early 1900s had seen Gertrude Winthrop continuing her social life in the United States whilst Lawrence had enlisted in the Northumberland Hussars, fought in the Boer War in 1900 – 1902 and then attended their summer camps lasting 10 days. By early 1904 he was again living at Little Shelford, Cambridgeshire, and had developed a keen interest in gardening becoming a Fellow of the Royal Horticultural Society and borrowing books on gardening from the RHS Library.

THREE

ARRIVING AT HIDCOTE IN 1907

The Hidcote Manor Estate which includes the village of Hidcote Bartrim with its dozen or so cottages occupies some 280 acres on a largely level plateau sheltered from the east winds some 600 feet above sea level in the northern outliers of the Cotswolds. It lies a mile east of Mickleton, 4 miles north of Chipping Campden and about 10 miles south of Stratford-upon-Avon on the Gloucestershire/Warwickshire border.

Hidcote is to the right of the clump of trees on the skyline to the right of the Church of St. James in Chipping Campden

The house, built in the seventeenth century as a farmhouse, was owned in the nineteenth century by the Freeman family. Captain William Thomas Freeman died on 4 December 1882 leaving all his real estate to his wife, Mary Webb Freeman. She died three years later on 21

September 1885 and left a bequest of *five hundred and twelve pounds or thereabouts which is owing to me on security of the security of a freehold estate at Hidcote in the County of Gloucester formerly the property of my late husband to my sister in law Mary Hannah Shekell Freeman.* This sister in law lived at The Martins in the High Street of Chipping Campden close to the twelfth century Market Hall and died eleven years later on 30 December 1906. In her will, she gave to her cousin Mary Franklin Hiron, *a yearly rent charge of one hundred pounds for her life to be charged upon and issuing out of all my lands and hereditaments known as the Hidcote Bartram* [sic] *Estate now in occupation of Mr John Tucker in the County of Gloucester* and also *subject to the said rent charge I give and bequeath my said Hidcote Bartram* [sic] *Estate unto the said John Tucker for all my estate and interest therein.* Probate was granted on 22 March 1907 with a gross value of £11,776-6-5.

John Tucker had been farming at Hidcote Bartrim since 1873. Within a couple of months of probate being granted, the estate was put up for auction. It was advertised in *The Times* on 22 June 1907 as a valuable freehold farm comprising some 287 acres and 34 perches to be sold by auction at the Noel Arms in Chipping Campden on Tuesday 2 July 1907 with possession on 29 September 1907 – Michaelmas day when most agricultural leases began and ended. The advertisement said that the farm would be sold together with the

> *very substantial and picturesque farm house, stone built, with, entrance hall, fine oak staircase, three sitting rooms, eight bed rooms, two box rooms, and usual offices, with lawns and large kitchen garden.*

It went on to note that

> *the farm is particularly healthy, being situate on a spur of the Cotswolds at an elevation of from 500 to 800 feet above sea level and from it extensive views of the counties of Warwick, Worcester and Gloucester can be obtained. Meets of the Warwickshire, North Cotswold and Haythrop* [sic] *Hounds are within easy distance, and the partridge shooting on the estate is good.*

Plan, Particulars, and Conditions of Sale

OF AN ESTATE AT

HIDCOTE BARTRIM,

GLOUCESTERSHIRE,

Three Miles from Campden and Long Marston Stations and Five Miles from Honeybourne Station, G.W.R.

VALUABLE AND DESIRABLE

FREEHOLD ESTATE.

HUTCHINGS AND DEER

Are favoured with instructions from the Owner to Sell by Auction, at the

Noel Arms Hotel, Chipping Campden,

On TUESDAY, the 2nd day of JULY, 1907,

At Five o'clock in the Afternoon, subject to the conditions of Sale annexed,

ALL THAT VALUABLE

FREEHOLD FARM,

COMPRISING

287a. 0r. 34p.

In 48 Enclosures, of which 134a. 3r. 19p. is Pasture, 141a. or. 26p. Arable, and 11a. or. 28p. Coppices, Woodland, Houses, and Buildings, together with the

VERY SUBSTANTIAL AND PICTURESQUE FARM HOUSE,

With Lawns and large Kitchen Garden,

Convenient and Substantial Farm Buildings,

AND

TEN COTTAGES

All of which is and has for 34 years past been in the occupation of Mr. John Tucker.

The Farm is in a good state of cultivation and will be sold with possession at Michaelmas next. Part of the Arable is good light land, the remainder being a rich loam with enough staple to grow corn of any description, or roots which can be eaten on.

The Pasture is excellent, and adapted for Dairying and Stock Rearing, and some of the enclosures will feed.

The Home and Farm are particularly healthy, being from 600 to 800 feet above sea level, and are approached and intersected by good roads.

There is an excellent bed of stone on the Property, with a Quarry which has been worked very remuneratively.

There are about 3 acres of good and productive Orcharding.

The House and Buildings are nearly all stone, with slated or tiled roofs in good condition.

The Property is well watered, the supply to the House and Buildings being laid on from the Hill.

Hidcote Manor sales particulars, 2 July 1907

The sales particulars described the residence as being very picturesque and the garden was described as consisting of *lawns in front and on the south side of the House, with fine shrubs and a nice Summer House, and a large and productive Kitchen Garden. Adjoining is a Tennis Court and small nut orchard.*

THE RESIDENCE

Is a very picturesque and substantial stone erection with a tiled roof, has a good elevation, and comprises Entrance Hall, with fine Oak Staircase, two Sitting Rooms, Office, Kitchen and Larder, 4 Bedrooms and 2 Dressing Rooms on the first floor, with 4 good Attics or Bedrooms and Box-room over. Large paved Court, Back Kitchen and Dairy with Granary over (Stone with Slated Roof), Washhouse, Coalhouse, and 2 w.c.'s.

There are Lawns in front and on the south side of the House, with fine shrubs and a nice Summer House, and a large and productive Kitchen Garden. Adjoining is a Tennis Lawn and small Nut Orchard.

Near the Court is the Nag Stable for 4 Horses with a good room over, and saddle room adjoining.

A plentiful supply of good water is laid on from the Hills to the House, Garden, and Buildings.

Extract from Sales Particulars for Hidcote Manor, 1907

These buildings are shown on the Ordnance Survey map of 1885.

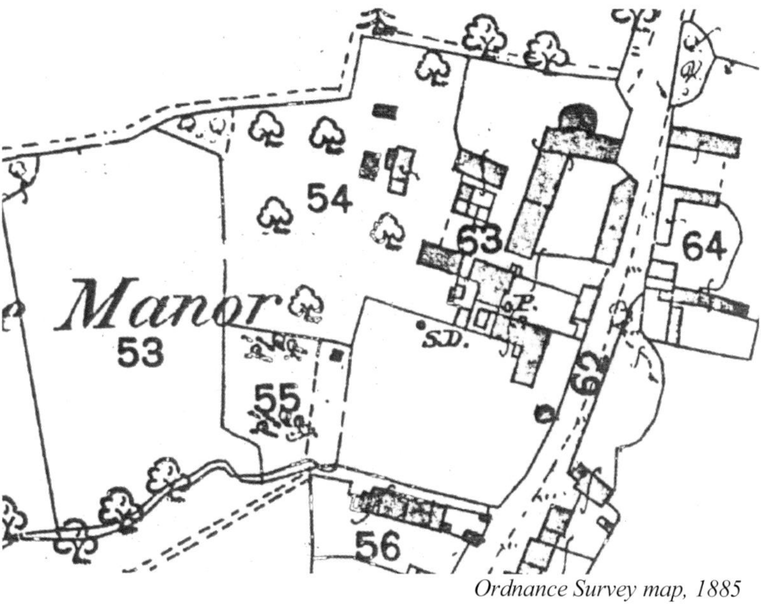

Ordnance Survey map, 1885

The sales particulars describe the various numbered areas as being:

63	House and Garden		1.125 acres
53	House Ground	Pasture	3.397 acres
54	Grass Bank	Pasture	1.098 acres
55	Paddock and Buildings		0.863 acres
56	Nut Orchard		0.228 acres
62	Road and Lawn		0.828 acres
64	Buildings		0.286 acres

Interestingly, the copy of the sales particulars deposited in the Gloucestershire Record Office by the solicitor, Oliver New, handling the sale for the vendor is annotated in pencil showing that the bidding was as follows:

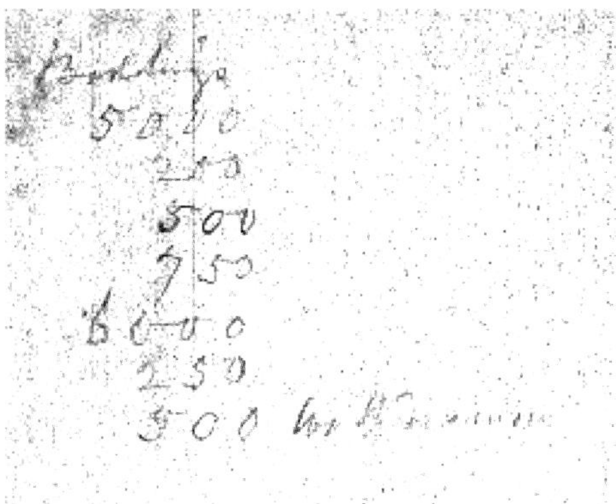

Annotations on Hidcote Sales Particulars

This shows Hidcote Manor was withdrawn from sale when bidding reached £6,500 (equivalent to about £580,000 today).

Three weeks later, on 23 July 1907 Gertrude Winthrop appointed her son, Lawrence Johnston, as her attorney to complete the purchase of Hidcote Manor Estate of 287 acres and thirty-four perches which Lawrence had contracted with John Tucker to purchase for the sum of seven thousand two hundred pounds (equivalent to about £650,000 today) a deposit of ten pounds having been paid. The Power of Attorney authorised Lawrence to enter into any mortgages necessary to complete the purchase. Interestingly, it gives Lawrence's address as being Little

Shelford in the County of Cambridge and Gertrude's signature is witnessed by E. W. Canham of Little Shelford, Valet. This confirms the information that Lawrence was living at Little Shelford from 1904 to 1907 and that his valet was E. W. Canham.

The reason why Gertrude appointed Lawrence as her agent was because she was about to sail to New York to resume her social life. She sailed on the *Kronprinz Wilhelm* a week later on Wednesday 31 July 1907 from Southampton accompanied by Ethel Boulton, aged 28, a god-daughter of Gertrude's, and a maid, Annie Sophia Mason, also aged 28. Lawrence's papers regarding his service in the Northumberland Hussars show Ethel's mother, Mrs. Boulton of Woolsthorpe, Findon, Surrey, as an alternative to his mother, Gertrude, as next of kin. A month later, on 8 September, the *New York Times* reports that Gertrude has given a lunch at the Malvern Hotel, Bar Harbor, Maine and that her 12 lady guests included Miss Boulton. A further entry in the *New York Times* in February 1908 notes that New Yorkers staying at the Ponce de Leon hotel, St. Augustine, Florida include Gertrude and Miss Boulton, her god-daughter. Then on 16 June 1908 Gertrude is listed as sailing today on the *Kronprinz Wilhelm* for Plymouth, showing that Gertrude stayed in America over the winter leaving Lawrence to move into the Manor House at Hidcote and possibly to have set in hand work on the extension. On this transatlantic crossing Gertrude was accompanied by Miss Hedwig von Lekow, a Hungarian friend, who received a bequest from Gertrude in her will almost 20 years later, and was the daughter of Baroness von Lekow.

The actual conveyance of the Hidcote Manor Estate took place in an Indenture dated 30 September 1907 (the day after the quarter day on which most rents involving farm land were due). The indenture is a complex document as it is between four parties – the first being the mortgagees of Hidcote Manor Estate to John Tucker; the second being Mary Franklin Hiron (who had been given an annuity of £100 based on an annual rent charge for the Hidcote Manor Estate by her cousin Mary Hannah Shekell Freeman); the third being John Tucker who had inherited the Hidcote Manor Estate from Mary Hannah Shekell Freeman; and the fourth being Gertrude Cleveland Winthrop. The indenture also specifies just how the purchase price of £7,200 for the Hidcote Manor Estate is to be paid: £4,900 goes directly to the mortgagees, £1,226–10–0 goes directly to Mary Franklin Hiron and the balance of £1,073–10–0 to John

Tucker. In a statutory declaration made by John Tucker on 2 October 1907, he declares that he is 63 years old and that he has for the past 34 years (since 1873) lived at Hidcote Bartrim and for the whole of that period has *occupied and farmed the Estate.* He declares that when he first arrived at Hidcote the estate belonged to Captain William Thomas Freeman to whom he paid his rent. When Captain Freeman died in 1882, the estate passed to his widow, Mrs. Mary Webb Freeman, and then on her death to Captain Freeman's mother, Mrs Priscilla Freeman, to whom he duly paid his rent. When she died in 1893, the estate passed to Miss Mary Hannah Shekell Freeman to whom John Tucker paid his rent until her death in 1906. Under her will the Hidcote Estate passed to John Tucker subject to the mortgage debt thereon. John Tucker declared that he had since her death paid no rent to anyone but had kept down the interest on the mortgage.

Gertrude also required a mortgage when she purchased the Hidcote Manor Estate. This was for the sum of £4,000 which was provided by Herbert Johnston Roper of Solihull and Bernard Philpin of Kington in the County of Hereford. Interest was payable of 4 per cent per year due in half-yearly payments on 1 October and 1 April. The mortgage was discharged in November 1913.

If a similar rate of interest (4 %) is assumed for the £4,900 mortgage on Hidcote whilst John Tucker was living there, this would have meant an annual payment of £196 which together with the annual rent charge of £100 to Mary Franklin Hiron would have come to a total each year of £296. It is possible to deduce that the estate had been withdrawn from the auction when bids reached £6,500 as the amount that would then have gone to John Tucker after paying of the mortgage of £4,900 and the annual rent charge of £1,226–10–0 would have been only £373–10–0, slightly more than his annual outgoings.

The reason why Gertrude and Lawrence came to Hidcote is not known. There are two possibilities – first that they were keen to be near the American community that was located in and around Broadway in the latter years of the 19th century and second that they were looking for a house near to Mark Fenwick of Abbotswood near Stow-on-the-Wold who Lawrence had come to know in the Northumberland Hussars or during his time in Northumberland. Mark had married Mary Sophia Clayton, the sister of George Saville Clayton, who was a fellow lieutenant with

Lawrence in the Northumberland Hussars in the 1900s. Of these, the more likely is a desire by Gertrude to be near the American community in Broadway, only 7 miles away from Hidcote Bartrim.

This community in Broadway had many American connections including Henry James, the author, who was born into a wealthy New York family in 1843 and moved to England in 1876 living first in London and then from 1897 at Lamb House, Rye, Sussex. From autumn 1885 he began to make annual visits to Broadway which he described as *the perfection of the old English rural tradition* that had become a gathering point for romantically inclined Americans in search of paradise. The first American to settle in Broadway was Francis David Millet, a painter, who arrived in 1884 and lived at Farnham House later moving to Russell House. A group of fellow artists gathered at Broadway with Ernest Abbey, another American painter, arriving in 1885. A close friend of Abbey was Alfred Parsons, the British landscape painter and garden designer who lived for some years in the USA, who also came to live at Broadway. Yet another American who joined the group in 1885 was the artist, John Singer Sargent. The annual Almanac for Evesham includes a section on Broadway which in 1905 said that Broadway was *one of the healthiest and most picturesque villages in England. It numbers among its residents personages distinguished in the artistic, literary, and dramatic world, and each year receives an increasing number of visitors from all parts of England, America, and the Colonies.*

Another American living at Broadway at this time was Mary Anderson de Navarro, the actress, who lived at Court Farm with her husband Antonio de Navarro. She wrote in her book *A Few More Memories* that when she arrived at Broadway she was astonished to find she knew many people in the neighbourhood: the Coventrys, the Elchos (Wemyss) of Stanway, Bruces, Gainsboroughs, Redesdales, Sir Arthur Blomfield, the Flowers, Hodgsons, and Liffords. She had met them all while she had been on the stage.

Mary Anderson in her book in a chapter entitled *A Few of Our Neighbours,* which was published in 1936 and thus covers many years at Broadway, noted that *Some of our neighbours have gardens which are of outstanding beauty.* She went on to mention that the garden at Batsford Park, had a European reputation and that the owners are old friends of ours. She goes on to say that Lawrence Johnston was another friend and the next paragraph says:

> Another famous garden is Lawrence Johnston's at Hidcote. The greater part of it is divided up into rooms, as it were, by yew hedges, each "room" containing a wonderful colour scheme. There is also a broad grass path, flanked by high yew hedges; the broad path dips at a gentle incline, then rises until it reaches two pillars with a delicately wrought iron gate—a gate which, standing on the skyline, seems as if it opens on the sky. Lawrence Johnston is a generous gardener and has given me many precious plants. My Italian friends regard Hidcote as the most beautiful garden they have seen in England. Its wonderful blending of colours and its somewhat formal, architectural character please them particularly.
>
> *Extract from Mary Anderson's* A Few More Memories

It therefore seems likely that Gertrude may well have met artists from the Broadway community or Mary Anderson either in New York or on the transatlantic liners. In any event, the Broadway community will have appeared attractive from the point of view of a lady who was active in New York society and wanted to continue such a life style in England.

When Gertrude and Lawrence came to Hidcote the description of the Manor House in the sales particulars said that it comprised:

> *Entrance Hall, with fine Oak Staircase, two Sitting Rooms, Office, Kitchen and Larder, 4 Bedrooms and 2 Dressing Rooms on the first floor, with 4 good Attics or Bedrooms and Boxroom over. Large paved Court, Back Kitchen and Dairy with Granary over (Stone with Slated roof), Washhouse, Coalhouse, and 2 w.c.'s.*

A photograph from that period shows the main entrance to the Manor House from the village road.

Hidcote Manor ca 1907

Another photograph shows the Courtyard and its buildings with the dairy and granary over on the left.

Hidcote Manor Courtyard & Barns ca 1907

A view from the village lane shows, to the left of the Manor House, the roofline of the dairy with granary over.

Hidcote Manor from Hidcote Bartrim village ca 1907

During the first few years after arriving at Hidcote, an extension was built which was linked to the house by a short connecting passage on the ground and first floors. This is shown in the plans of the house that date from 1913. These plans show that the original Manor House had, on the ground floor, a Drawing Room to the left of the original front door, a Sitting Room to the right and a Dining Room, sometimes referred to as the Hall, some 29 feet long by 16 feet 3" wide with a door to the Courtyard. On the first floor, there were three bedrooms and bathroom whilst on the top floor were two bedrooms and a tank room as well as a smaller room 11feet by 9 feet. Linked by a passage on the ground floor was the Kitchen with a bay window, a room for the Housekeeper 11 feet by 10 feet 9", a Servants Hall some 19 feet 6" long by 12 feet 6", a Scullery, Larder, Dairy and Laundry. On the first floor, linked by a passage from one of the bedrooms were three further bedrooms, a linen closet and bathroom. On the second floor, reached by stairs from the first floor of the extension, were three bedrooms, a box room and a servant's bathroom. There were consequently now a total of eleven bedrooms of which five were on the top floor and were probably for servants.

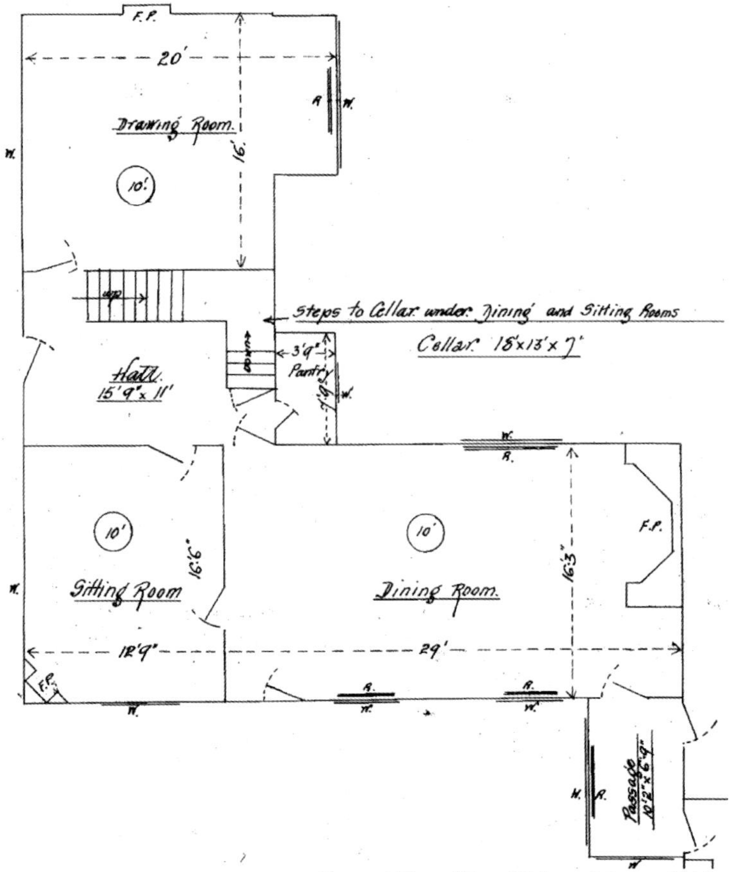

Ground Floor Plan, Hidcote Manor, 1913

The garden was described in the sales particulars as:

There are lawns in front and on the south side of the House, with fine shrubs and a nice Summer House, and a large and productive Kitchen Garden. Adjoining is a Tennis Court and small nut orchard.

The plan forming part of the 1907 conveyance is not very clear in relation to the gardens but it does show and name the various pieces of land near the Manor House:

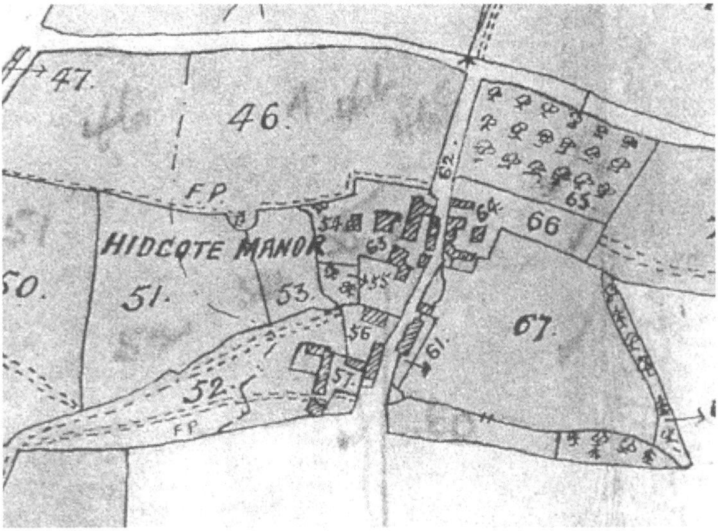

Hidcote Manor plan, 1907 Conveyance

The following are described in the Conveyance as follows:

63	House and Garden
46	Oat Piece
62	Road and Lawn
56	Nut Orchard
53	Grass Bank
52	House Ground
55	Paddock and buildings

There was little shelter from the prevailing winds as Hidcote Manor was surrounded by arable fields and pastures to the south, west and north. There was an orchard to the north-east across the village road and a few trees along some of the field boundaries. In the first published account of the garden some 20 years later in *Country Life*, Avray Tipping described the situation prior to Lawrence Johnston's arrival as having on the east side *a rectangular space lying between road and door, while on the south side a sloping lawn with a cedar tree and some flower beds formed a small pleasure garden, enclosed from the utilitarian sections and the fields. The general slope was southward, while westward was a modest but immediate rise, eastward a slight dip before a sharp ascent, and north more or less level ground, so that only buildings and trees could give protection.*

A photograph from the early 1900s shows the village of Hidcote Bartrim – the wall of the Manor House garden and the cottages on the right are unchanged today.

Hidcote Bartrim village in the early 1900s

A photograph of the garden looking towards the Cedar of Lebanon from the early 1900s predates the extension to the house.

Old Garden looking towards Cedar in early 1900s

The trellis work to the right of the Cedar of Lebanon suggests that, as Tipping said later, the ground is sloping southwards towards the wall round the Old Garden. No other photographs have yet emerged that show the garden before Lawrence Johnston and Gertrude Winthrop arrived. The photograph (page 67) from Hidcote Bartrim village showing the Manor House before the extension was built shows a tall pine tree in the part of the garden that is today the Maple Garden.

This then was the tapestry on which Lawrence Johnston set to work to create the garden at Hidcote. This photograph shows Lawrence with two of his Border Terriers shortly after he arrived at Hidcote:

Lawrence Johnston at Hidcote early 1910s

FOUR

CREATING HIDCOTE MANOR GARDEN 1907 – 1914

In creating the garden at Hidcote, it is evident that Lawrence Johnston was influenced by the ideas put forward by Thomas Mawson in his book *The Art and Craft of Garden Making* which appeared in its first edition in 1900 and then reappeared in subsequent editions in 1901, 1907, 1912 and 1926. This was a book that Lawrence had borrowed from the RHS Lindley Library in 1905 and borrowed it again in 1911.

Thomas Mawson The Art and Craft of Garden Making

Mawson describes the styles of laying out gardens identifying what is commendable in each and how these commendable features may be adapted to modern requirements. Successive chapters deal first with the selection of a site and then how lawns should be laid out, terrace and flower gardens created, hedges and fences, summer houses, water gardens and ponds as well as conservatories and kitchen gardens. There are many echoes at Hidcote of what is written in *The Art and Craft of Garden Making.* Thus Mawson notes that "*Repton recommended formality near the house, merging into the natural by degrees, so as to attach the house by imperceptible gradations to the general landscape*" thus making the point that the garden is the link which connects the house and the landscape – just as Lawrence's garden does at Hidcote. Mawson goes on to say that:

> *"The arrangement should suggest a series of apartments rather than a panorama which can be grasped in one view: art is well directed in arousing curiosity, always inviting further exploration, to be rewarded with new but never a final discovery."*

Hidcote is well known for its many garden rooms or apartments – ranging from the Maple Garden through to the Pillar Garden.

Work started on the creation of the garden soon after Lawrence arrived at Hidcote. There are no records showing who worked on the garden although deductions can be drawn from the census records for 1901 and 1911, recording all living in Hidcote Bartrim, and the annual electoral registers, showing which men were living in a property rated at £5 or more. These show that George Bennett, Thomas Handy, Alfred Holtom, Alfred Hughes, Thomas and William Newman, Thomas Stafford and William Pearce, along with his sons, Edward and Harry, were living at Hidcote Bartrim as was Ernest Daniels. It is known that Edward Pearce and his brother, Harry Pearce, the sons of William Pearce, both worked on the garden as did members of the Bennett and Hughes families. Ernest Daniels was the chauffeur.

Lawrence created the garden at Hidcote in phases with the first one during the years from 1907 until World War I began in 1914. In this phase, Lawrence changed the main entrance into the house to the one from the courtyard into the north side of the house, and on the east side of the house removed the steps up from the road, raised the wall separating the garden from the village road and decorated it with urns as can be seen in a recent photograph. He created a cobbled parterre garden some 19 feet wide – the East Court Garden – behind this wall.

The wall between the village road and the East Court Garden

However, in raising the wall between the village road and the East Court Garden, Lawrence clearly had an overall scheme in mind. This was to open up a vista running directly to the east that would be visible for those emerging from the Manor House into the East Court Garden. This vista would lead the eye through an avenue of trees to a statue at the end.

Consequently the wall that had been built between the village road and the garden was lowered in the centre between the urns so as to keep open the view from the door of the Manor House to the east across the ha-ha on the far side of the lawn over the road and up the Lime Avenue to the statue of Hercules. The edges of the avenue are framed by further similar urns on lower walls that define each end of the ha-ha. The sense of proportion shows the excellent eye that Lawrence had for both location and proportion which is shown throughout the garden at Hidcote.

View to the East up the Lime Avenue

This recent view to the east shows the pillars, with similar urns to those on the wall, on the lower wall at each end of the ha-ha that frame the view up the Lime Avenue.

To the south of the house was a large Cedar of Lebanon which along with the walls to the Old Garden are all that remain of the garden that was at Hidcote before Lawrence arrived. In developing what is now known as the Old Garden with its two apartments or garden rooms – the Maple Garden and White Garden at a level below that of the Cedar Lawn – Lawrence was again taking note of advice given by Mawson that:

> *"Having assimilated the natural and striking features of the site, study next its sectional lines and adopt the course which most frequently attains the happiest results. Working outwards from the house, dispose the terraces as the falls of the land allow, and the height of the house demands, according as the original disposition of the land suggests."*

Mawson goes on to note that *"every landscape architect whose work has obtained recognition agrees that, in all but exceptional cases, in order to give a proper connection between house and garden, a formal arrangement near the house is essential."* In designing a garden, Mawson says that:

> *"attention would first be directed to discovering and framing those features visible from it which have in them the elements of the picturesque, or which in any way give character and individuality to the site. "Nothing" says Sedding "is prettier than a vista through the smooth-shaven green alley or an archway framing a view of the countryside beyond," and "it is for the creation of such effects that the designer must aim in the arrangements of his terraces and particularly their steps and the placing of seats, arbours and bastions so as to emphasise them when created, at the same time taking care that the balance and symmetry of the scheme as a whole are not endangered in the treatment of individual features."*

He goes on to make the point that gates and fences *"ought to be arranged to enhance the beauty of the grounds they enclose or partition"* and that *"Gates may advantageously mark the end of a vista"* something that Lawrence was to use to excellent effect at Hidcote. Initially there was the gate at the western boundary of the Old Garden and then later, at the end of both of the long axes – westward from the Cedar Lawn through the Red Borders and the Gazebos to Heaven's Gate and the other axis running southwards from the Gazebos to the end of the Long Walk. Gates are again used to good effect in the northward vista running up the Beech Avenue to the north of the Great Lawn, now generally known as the Theatre Lawn.

In the Old Garden itself, the first two garden rooms – the Maple Garden at the eastern end of the garden together with the adjacent White Garden on the western side of the Maple Garden – were created at a level of a meter (3.5 feet) below that of the Cedar Lawn. Then, outside the gate, to the west of the Cedar of Lebanon, which provides the border of the Old Garden, the Circle was created with an axis running south first down through an intervening garden room – the Fuchsia Garden – and then down to the Bathing Pool Garden and then up steps to the Green Circle. Running west from the Circle ran the Red Borders then up steps between the Gazebos and west to Heaven's Gate. This was a continuation of the axis through the Old Garden under the Cedar of Lebanon to the gateway into the Circle and on between the Red Borders. The location of these early parts of the garden are shown on the annotated 1944 chain plan.

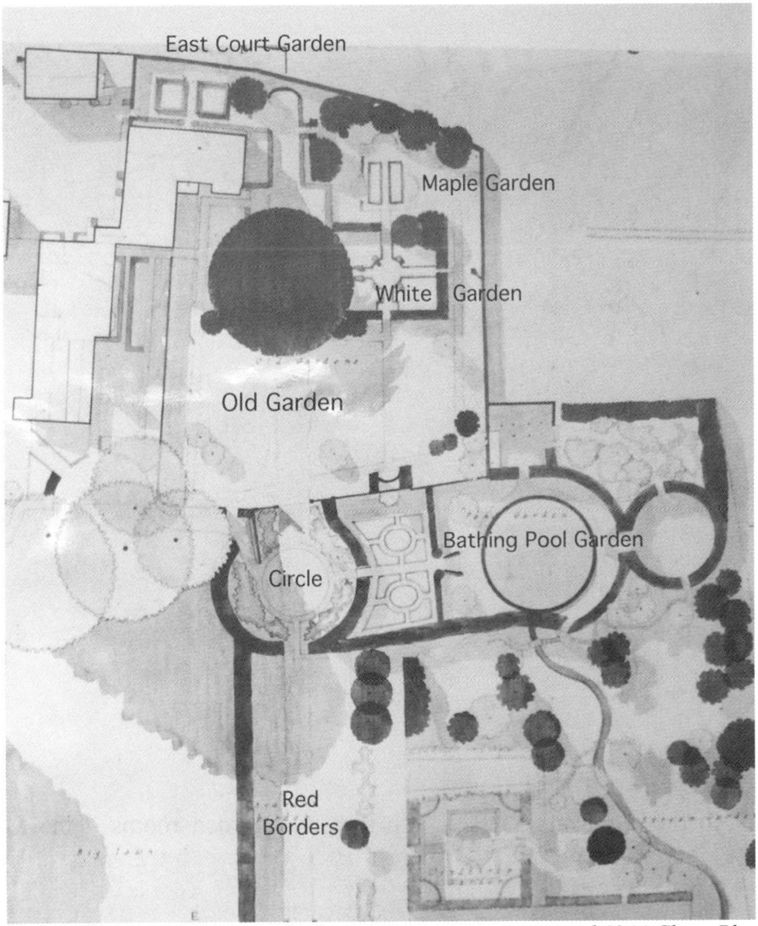

Annotated 1944 Chain Plan

An early photograph showed the sunken Maple Garden with its ground level beds with several large tubs on the surrounding paving slabs.

Maple Garden ca 1910

There are few mentions of the Maple Garden in the early accounts of the garden in the 1930s although Russell Page describes the central axis thus:

> *A pleasantly shaped, but not enormous, cedar tree is the starting point for a long vista which runs from south to north right through the garden. This vista consists of a wide grass path linking various gardens divided from each other by hedges.*

and goes on to add that:

> *To the west of this main garden vista lies a whole series of small enclosures separated by hedges for shelter and surprise. Below the cedar tree are gardens specially for paeonies and phlox, and one for rare lime-hating subjects.*

In the early years, the adjacent White Garden had a central squat sundial surrounded by a circle of grass and with paving slabs on the two axes. The extension to the house was in place but there were no topiary birds at the corners of the four beds in the White Garden, nor was there any hedge between the Cedar Lawn and the White Garden.

White Garden ca. 1910

The squat sundial was subsequently replaced by a slender sundial made from a pilaster from the second bridge across the Thames at Richmond; this sundial is today in Mrs. Winthrop's Garden. At the same time as the sundial was replaced, small topiary birds were planted at the corners of the four beds of the White Garden and two round topiary shapes were created on either side of the steps down from the Cedar Lawn. The bench seen under the Cedar is also visible in photographs of Gertrude Winthrop and of her butler and gardeners at Hidcote.

White Garden looking away from Cedar Lawn ca 1913

A photograph in the other direction shows there is still no hedge to the Cedar Lawn but is otherwise very similar in respect of the topiary birds.

White Garden towards the Manor House ca 1913

A photograph looking towards the west from under the Cedar of Lebanon shows young topiary birds and the vestiges of a hedge around the Cedar Lawn with a narrower and taller gate than that later inserted by Lawrence in the wall of the Old Garden. Lawrence was thus reflecting Mawson's advice that *Gates may advantageously mark the end of a vista.*

View to west from Cedar Lawn towards Gate of the Old Garden, ca 1912

Further west, outside the Old Garden, a start had been made on a vista to the south through what is now known as the Bathing Pool Garden which then had a small ground level pool. This is approached from the Circle by a few steps to the south down into the Fuchsia Garden some two feet below the level of the Circle and then by some further steps descending 3 feet to the Bathing Pool Garden and then steps up to the Green Circle.

An early photograph shows the ground level pool with five paths radiating in various directions from the southern side of the pool and unplanted beds to the east and west of the path leading to the steps up to the Fuchsia Garden. Although hard to discern, topiary birds can be made out on either side of the top of these steps. As to the Circle, pergolas can be seen providing an early boundary between the Fuchsia Garden and the Circle. In introducing the pond, Lawrence was following Mawson's advice that *it may be questioned whether they* [gardens] *are complete without it* [water], *if only a small pond, reflecting and blending, in thousands of beautiful ways, the hues of flowers, foliage and sky, at the bidding of every passing breeze.*

Bathing Pool Garden looking towards Manor House ca 1912

A further view at about the same time shows the Fuchsia Garden looking towards the Circle, the wall and the narrow gate to the Old Garden and the extension to the Manor House.

Fuchsia Garden looking towards the Manor House ca 1912

Another photograph of the Bathing Pool Garden at about the same time but looking southwards shows the Fuchsia Garden, the topiary birds and, beyond the ground level pool, the yews planted on top of the stone wall to provide shelter to the pool, and beyond, the Green Circle.

Looking towards the Bathing Pool Garden ca 1912

These photographs showing the Bathing Pool Garden and its adjacent garden areas are fascinating as they show the foresight that Lawrence had in planning his garden. The steps down from the Circle to the Fuchsia Garden, and then from the Fuchsia Garden down to the Bathing Pool Garden and then up from the Bathing Pool Garden to the Green Circle are all in place in these early photographs and demonstrate the planning and sense of proportion that enabled Lawrence to envisage the final creation and thus where to locate the steps at the outset. He was clearly following Mawson's guidance that *the natural contours of the land must be incorporated into the scheme.* The pergolas in these views of the Bathing Pool Garden and the Fuchsia Garden show that he was again following Mawson's advice that *In a new garden, where shade is difficult to obtain, a pergola is invaluable, as it can be quickly covered with foliage and flowers and so prove very useful until the planted trees have grown.*

Another slightly later view towards the Bathing Pool Garden shows how the topiary birds and the yews round the pool have grown.

Looking towards the Bathing Pool Garden ca 1914

During this same period, work had commenced on the vista to the west through the Red Borders and to build the Gazebos. The topiary birds by the Bathing Pool Garden are visible and the Italian Shelter on the extreme right hand side is under construction. The trellis work around the Circle is evident together with the narrow gate into the Old Garden.

Red Borders looking toward the Old Garden ca 1912

Lawrence Johnston's horticultural ability is underlined by the fact that during these years whilst he created the garden at Hidcote, he was also able to win an Award of Merit for the Hidcote strain of *Primula pulverulenta* at the RHS show held in Vincent Square, London on 6 June 1911.

> To *Primula pulverulenta* 'Hidcote strain' (votes, unanimous), from L. W. Johnston, Esq., Hidcote Manor, Campden, Glos. The strain has the habit of the type, but the flowers are mostly of a light rose colour with a greenish yellow centre.
>
> *Proceedings Royal Horticultural Society, JRHS, 37, 1911-2, p. cxxxvi*

The account in the Times the following day provides further information:

> Mr. L. Johnston (Campden, Glos.) won an award with the Hidcote strain of primula pulverulenta. Of the many new Chinese primroses none has spread from garden to garden so rapidly as pulverulenta, but hitherto the typical crimson flowers only were known. The Hidcote strain adds the soft pinks from primula Japonica, at the same time retaining the mealy stem and other characteristics of pulverulenta.
>
> *The Times, 7 June 1911*

In addition to Lawrence's activity in the garden, he continued to serve with the Northumberland Hussars and attended their summer camps. He was promoted to Lieutenant on 27 May 1905 and was at the summer camp at Walwick Grange in June 1907 and again in June 1908. He was promoted to Captain on 28 May 1909 and then to Major on 7 June 1913. In May 1914 he was at the annual camp at Farnley Camp (Otley, Yorkshire) and then when mobilization took place for the Great War he was at Gosforth Park, Newcastle-upon-Tyne, commanding the 'A' Squadron of the Northumberland Hussars. On 11 September, the Northumberland Hussars travelled by train to Lyndhust, Hants with Major Johnston being the Transport Officer, and on 5 October 1914 sailed for Belgium.

Meanwhile, Mrs. Winthrop was continuing her social life. As noted in the previous chapter, Gertrude had sailed from New York on 16 June 1908 on the *S. S. Kronprinz Wilhelm* for Plymouth and is listed among the First Class passengers who arrived on 23 June. The papers attached to her U.S. passport application in 1917 make it clear that following her arrival in 1908 she made one visit to the United States in 1912 for a few months.

There is little information about what she was doing between 1908 and 1912. There is mention in a local diary of a lunch party at Hidcote on Tuesday 30 March 1910 given by Gertrude and Lawrence at which the Rev. Jackson, Vicar of Chipping Campden, and his wife, the Paleys, May Bruce, the Ashwins of Bretforton and Mr. Wrigley were present. In addition, it is likely that Gertrude will have travelled abroad and may have stayed in London for some time as some of her correspondence was sent from the Hans Crescent Hotel, Knightsbridge which was popular with London society – and its manager witnessed a codicil to Gertrude's will in later years.

1912 was to be an eventful year with a notice in the New York Times on 27 March saying that Elliott Johnston, son of Mrs. C.F. Winthrop, had died suddenly on Saturday, 16 March, in Los Angeles, California. The funeral service was to be at St. Peter's Church, Westchester – the same church at which Gertrude had married Lawrence Johnston's father in 1870 and where her parents are buried – on Thursday, 28 March, at 3.30 pm followed by interment at the Woodlawn Cemetery, Bronx, New York. The *Orange County Tribune* in its issue of 20 March 1912 contained an item on the front page headed *Severely Burned in Explosion of Gas: Elliott Johnson* [sic] *of Delaware Union Victim of Accident.* This article reported that:

> *Elliott Johnson* [sic], *aged about 40 years, was very severely burned as the result of an explosion of gas in the bunkhouse on the Delaware Union base Friday morning about 4 o'clock. The exact cause of the accident not known as there was no witnesses near the house at the time. Mr. Johnson was seen to run out of the house, with his clothes ablaze, crazed by pain.*

The article went on that *It is thought that during the early part of the morning that he was awakened by the sound of gas escaping into his room. Natural gas from the wells on the base was piped into his house and it is probable that the rubber tube connecting with the gas stove under high pressure "blew off".* It then added that *It is likely that when he entered he lighted a match and air and gas mixing ignited.* It concluded by saying:

He was badly burned about the head, arms and body and his chances of recovery were pronounced by the attending physicians as slight.

Johnson was related to the Waterburys of New York who are part owners of the oil property.

The *Los Angeles Times* on 17 March included an item dated 16 March reporting that:

Elliott Johnson [sic], *who was severely burned in a gas explosion in his bunkhouse on the lease of the Delaware Union Oil Company in Brea Canyon yesterday morning, died today at the Anaheim Sanitarium. Johnson was a relative of the Waterburys of New York, who own a large interest in the Delaware Union and was himself well to do.*

Gertrude and Lawrence along with Hedwig von Lekow and a maid Jessie McElroy all sailed from Southampton on 29 May the same year on the *S.S. Kronprinzessen Cecilie* arriving in New York on 4 June. They were in New York for some six weeks as Gertrude, again accompanied by Hedwig von Lekow, is listed among the First Class passengers who arrived from New York on the *S.S. Kronprinz Wilhelm* at Plymouth on 22 July. Lawrence had returned a month earlier, arriving on 22 June at Plymouth aboard the *S.S. Olympic.*

The death of her son, Elliott, seems to have somewhat unsettled Gertrude because by December 1912 she was considering selling Hidcote. A letter of 23 December to Messrs. Bruton Knowles & Co, a firm of estate agents and auctioneers in Gloucester, from Edward J. Tozer, a solicitor in Teignmouth, says that:

I have mentioned the name of your firm to a well to do American lady, a Mrs Winthrop of Hidcote Manor, Campden, Glos., who wants to get in touch with a thoroughly reliable firm in connection with the contemplated sale of Hidcote. I may tell you I have only met the lady once personally, but I know of her through my sister in law who stays with her frequently.

The letter goes on to add that:

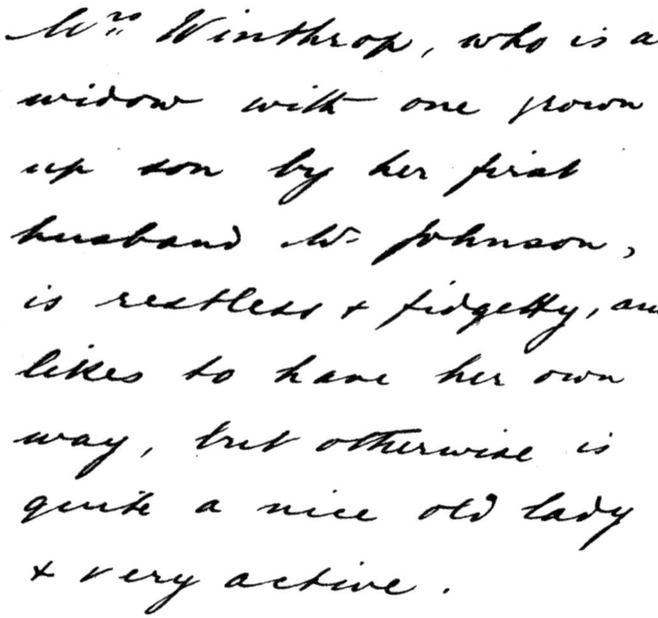
Mrs Winthrop, who is a widow with one grown up son by her first husband Mr Johnson, is restless & fidgetty, and likes to have her own way, but otherwise is quite a nice old lady & very active.

Edward Tozer letter, 23 December 1912

It continues by saying that

> *I believe she has spent a great deal of money on Hidcote, & that her idea of leaving is due to reasons of health which are attributed to the hardness of the water, soft water being almost a mania with her.*

The matter is taken forward by a letter on 7 February 1913 from Geoffrey & Oliver New, who were Gertrude's solicitors in Chipping Campden, to Messrs. Bruton, Knowles & Co. saying:

> *Mrs. Winthrop, who is a client of ours, wishes us to write to you in reference to her property, Hidcote Manor, and land adjoining, the whole comprising about 287 acres. She has been living there for some few years, and finding that the climate does not suit her she would be disposed, provided she could get a good price for the property, to sell it.*

The letter goes on to add that:

We may say that the property, together with the extensive additions and alterations which have been made, has cost our client about £16,000.

It is annotated in pencil that *Our client is shortly gone abroad and the matter is entirely in our hands.*

There is then an exchange of letters and telegrams which lead to a letter on 21 February from Oliver New to Bruton Knowles saying *We have seen our client today and find she is leaving England on 3rd March for 8 or 10 weeks and would like to see you with our Mr O. H. New on Thursday next if possible at her house.* The visit took place and subsequently Gertrude wrote to Mr. Bruton on 27 February saying that her bailiff has told her that the rates and taxes amount to £48 a year on the house and farm buildings and that the tithes are £10 a year. The letter goes on to say:

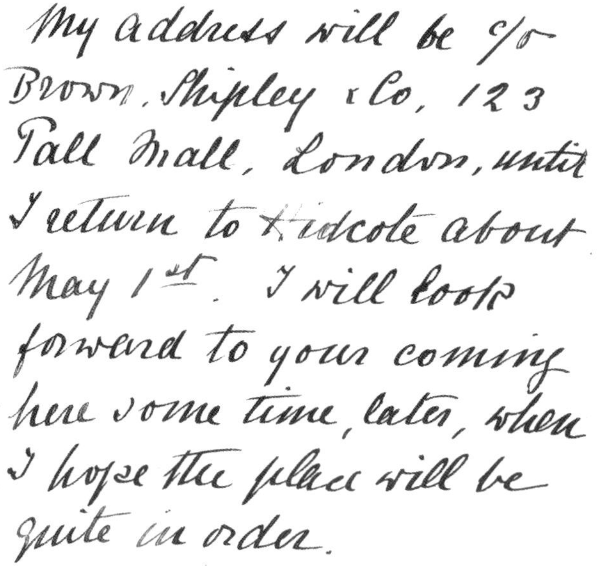
My address will be c/o Brown, Shipley & Co, 123 Pall Mall, London, until I return to Hidcote about May 1st. I will look forward to your coming here some time, later, when I hope the place will be quite in order.

Mrs. Winthrop letter, 27 February 1913

Brown, Shipley & Co. were a private banking company set up by Brown Bros., a bank in Baltimore, Maryland, who opened an office at 123 Pall Mall in 1900 to handle the many American visitors carrying their Letters of Credit.

Her letter of 27 February continues:

> *I can then also try to have some good photographs taken of the house.*
>
> *The Campden motor will, of course, be charged to my account. Thank you for wanting to arrange it otherwise. When you come here again, you must let me know so that I can send for you.*

These photographs of the house have not yet appeared. The note about the cost of the Campden motor and arranging to have Mr. Bruton met on his next visit give an insight into Gertrude's way of life.

Bruton Knowles wrote to Oliver New on 1 March enclosing Gertrude's letter of 27 February and saying:

> *I have been thinking the matter over and I feel very strongly that it would be far better to withdraw the property from the market until Mrs Winthrop returns. The alterations will then probably be completed and the whole place will have a far better appearance. I notice that Messrs. Curtis & Henson are advertising it in 'Country Life' which I think is unwise at present. I would suggest that you withdraw the property from both our books.*

The Curtis & Henson advertisement in Country Life on 1 March 1913 described Hidcote without actually naming the property.

COTSWOLD COUNTRY (in A MOST DELIGHTFUL SITUATION, 500FT. ABOVE SEA LEVEL, two miles from a station).—A CHOICE OLD MANOR HOUSE, the subject of extensive alterations and additions during recent years, now in excellent order throughout. FOR SALE with about 300 acres of good land. HUNTING with HEYTHROP.—Strongly recommended from personal inspection.—CURTIS & HENSON.

Country Life advertisement, 1 March 1913

Oliver New replied on 3 March agreeing with Bruton Knowles and withdrawing Hidcote from the their books. Further correspondence related to the preparation of a plan of Hidcote. Then on 22 May Oliver New wrote to Bruton Knowles to say:

> *Referring to our recent interview with you, we write to tell you that our client Mrs Winthrop has now returned from the Continent, and that we saw her yesterday. She appears to have made up her mind to dispose of the Hidcote property, but is in no particular hurry, and is at present looking round to see what else she can find to suit her. We understand that she is going somewhere in the neighbourhood of Bath next week to inspect some property there, and she wishes us to ask you to be on the lookout for anything you think likely to suit her, and to let us have particulars. We rather think she would prefer to rent a place, at any rate at first, so as to see whether it suited her or not, and would probably like an option of purchase if it could be obtained. We think you know that the great essentials are soft water and a mild climate.*

There is no record of any further correspondence until the end of the year when a letter of 31 December from Oliver New to Bruton Knowles says *We doubt however whether she is now likely to desire to sell.* A reply on 5 January 1914 makes it clear that Bruton Knowles *are not at all sanguine that business will result.* And there the matter rested. Whether this was because Gertrude had become reconciled to the death of her son, Elliott, or because of the growing war clouds in Europe, we do not know. Hidcote was not sold and Lawrence was fortunately able to continue to create the garden.

FIVE

THE GREAT WAR YEARS

1914 saw the storm clouds of war gathering over Europe. The Northumberland Hussars, with Lawrence Johnston as a Major, were mobilized at Gosforth Park, Newcastle upon Tyne. Lawrence was in command of the "A" Squadron and the regimental orders on 19 August show that he was signing the orders for that squadron.

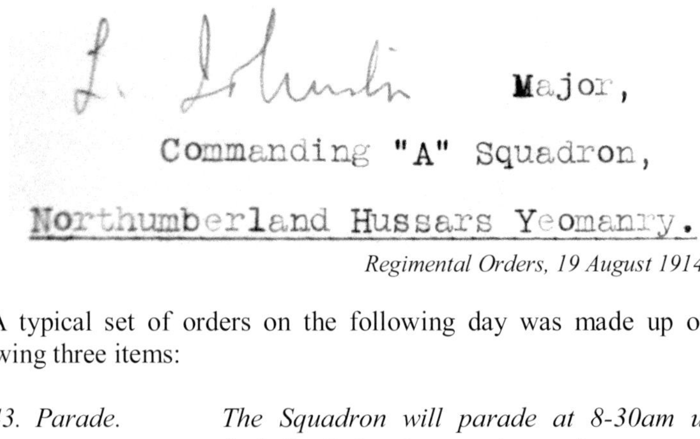
Major,
Commanding "A" Squadron,
Northumberland Hussars Yeomanry.

Regimental Orders, 19 August 1914

A typical set of orders on the following day was made up of the following three items:

43. Parade.	*The Squadron will parade at 8-30am under 2nd. Lt. Laing in watering order, one man to two horses.*
44. Leave.	*Leave will be granted to 20 per cent of the men of the Squadron from 6-0pm to 9-0pm until further orders.*
45. Concert.	*A Concert will be held in the Luncheon Room beneath the Grand Stand on Friday at 7-30 pm.*

Members of the Regiment will be allowed to take their civilian friends. Entrance by front of the Grand Stand.

These Regimental Orders continue until 11 September 1914. At 5pm on that day, the order came for the Regiment to start the next day by train for Lyndhurst, Hants. Major Cookson was in command of the Regiment and Major Johnston was Transport Officer. Then on 5 October, the Northumberland Hussars Yeomanry sailed on the *S.S. Minneapolis* from Southampton across the Channel to Zeebrugge, Belgium. They then marched to Bruges en route to join the Seventh Division as Divisional Cavalry and eventually entered Ypres on 14 October. In the subsequent "*First Battle of Ypres, 1914*", the History of the Northumberland (Hussars) Yeomanry records for the 23 October 1914:

This was a day of severe fighting for the regiment, a day on which one of the most dangerous thrusts of the enemy was successfully foiled. Under increasing pressure the infantry had been forced to give ground, and it was just at the moment when the gap was ominously widening that the regiment, waiting in reserve, was called in to assist. .... Here we remained several hours under very heavy rifle and shell fire, unable to retaliate very effectively, owing to the poorness of the field of fire. But these gallant riflemen stuck to it, their crisp sharp fire orders never seeming to falter. Then came the crowning incident of the day. A line of Scots Guards suddenly rose to the order of "Come on, the Scots Guards!" echoed by Major Sidney's "Come on, Northumberland Hussars!" and together guards and Hussars charged against a swaying mass of grey figures and finally drove them over the hill. Our casualties, considering the severity of the fighting, were relatively light, but among the wounded were Major Johnston and Lieutenant Laing.... It was a rough piece of fighting, and the regiment had acquitted itself most creditably.

Lawrence Johnston's army records show that his wounds meant that he left his unit on 23 October 1914 and embarked on the *S.S. St. Patrick* at Boulogne on 30 October arriving at Southampton on the following day. A Medical Board on 3 November recorded that *he is suffering from a shrapnel bullet wound of the right side of the chest* and categorised the wound as being *'Severe but not permanent'* and estimated that he would be incapacitated for a period of ten weeks from the date of his injury. The report of the Medical Board is annotated *9 Grosvenor Gardens, S.W.* which was the address of the King Edward VII Hospital for Officers.

The *Illustrated Chronicle* issued in Newcastle contained an item in its issue of 27 October 1914 headed:

NEWS OF THE HUSSARS

Northumberland Regiment "Did Splendidly."

NONE KILLED.

Six Officers Wounded in Yeomanry's First Real Taste of the War.

Illustrated Chronicle, 27 October 1914

It also reported that Lord Ridley, the commanding officer of the regiment, had received a telegram the previous day saying Major Johnston and five other officers had been wounded but *all going on well.* The same telegram said the *Regiment did splendidly*. The item also included a section headed *Who's Who* with a short paragraph on each officer:

Who's Who.

Major L. Johnston served in the South African War, from 1900 to 1902, in the ranks of the Imperial Yeomanry. He took part in the operations in Cape Colony, north of Orange River, including the action at Ruidam, and also in the fighting in Orange River Colony and the Transvaal. He received the Queen's medal with three clasps and the King's medal with two clasps.

Illustrated Chronicle, 27 October 1914

On another page, the same issue of the *Illustrated Chronicle* had a photograph of Major Johnston:

Illustrated Chronicle, 27 October 1914

The *Roll of Honour* published in *The Times* on 7 November 1914 included amongst the wounded:

Johnston, Maj. L., Northumberland Hussars

With Lawrence Johnston in the King Edward VII Hospital for Officers in Grosvenor Square, it is interesting to find that the records of the Lindley Library of the Royal Horticultural Society show that on 30

November 1914 he began borrowing books on gardening and on plant hunting. Vincent Square, where the Lindley Library is located, is about half a mile – or only some ten minutes walk – from Grosvenor Gardens thus facilitating the loans. All these loans were signed out for Lawrence and not by him – on occasion the rank was shown incorrectly as *Captain* instead of as *Major*. Some of these loans were signed for by William Brown, the butler at Hidcote. It is thus possible to envisage Lawrence sitting up in bed recovering from his wound between November 1914 and March 1915 and reading these books:

Date of loan	Date of return	Name of Book
30 Nov 1914	1 Dec 3 Dec 1 Dec	Castle, *Orchids* Millican, *Orchid Hunter* Wright, W.P., *Alpines & Rock Garden*
1 Dec 1914	7 Dec 7 Dec 7 Dec	Malby, *Oberland & Valais* Trevor-Bathye, *Crete* Ward, *Land of the Blue Poppy*
7 Dec 1914	29 Dec 29 Dec 29 Dec	Spruce, *Notes of a Botanist* (2 vols) Cecil, *History of Gardening* Annesley, *Beautiful Trees*
29 Dec 1914	12 Jan 29 Dec	Godfrey, *Gardens in the Making* Wilson, E. H., *A Naturalist in Western China, 2 vols*
29 Dec 1914	12 Jan 12 Jan	Hooker, J. D., *Himalaya Journals 1905* Du Cane, *Gardens and Flowers of Madeira*
12 Jan 1915	2 Feb 4 Feb 2 Feb	Elgood & Jekyll, *Some English Gardens* Sedding, *Garden Craft Old & New* Triggs, *Garden Craft*
2 Feb	4 Feb	Elwes, *Trees vol. 1 - 3*
4 Feb	5 Feb	Elwes, *Trees of Great Britain & Ireland vol 4-6*
5 Feb	17 Feb 17 Feb	Wright, *Fruit Growers Guide* Bean, *Royal Gardens Kew*
17 Feb	20 Feb 20 Feb 20 Feb	Davidson, *Unheated Greenhouses* Hasluck, *Greenhouse Construction* May, *Greenhouse Management*
26 Feb	25 Mar 25 Mar 25 Mar	Bunyard & Thomas, *The Front Gardens* Rowles, *Gardens under Glass* Trevena, *Adventures among wild flowers*

These books included accounts of plant hunters – notably Albert Millican's book *Travels and Adventures of an Orchid Hunter* (1891), an account of canoe and camp life in Colombia, while collecting orchids in the Northern Andes; Frank Kingdon-Ward's book *The Land of the Blue Poppy* (1913) about his expedition in 1911 to the north of Yunnan and South-Eastern Tibet to collect plants for the horticultural firm of Bees Ltd Liverpool; Ernest H. Wilson's book *A Naturalist in Western China* (1913) on eleven years travel and exploration in China; and Joseph Dalton Hooker's book *Himalayan Journals* (2nd edition 1905) which are the notes of a naturalist in Bengal, the Sikkim and Nepal Himalyas, and the Khasia Mountains. Ernest Wilson was the plant hunter who was born in 1876 in Chipping Campden some four miles from Hidcote Manor garden.

The books borrowed by Lawrence at this time included books on garden design including Walter H. Godfrey's book *Gardens in the Making* (1914) which describes various features of gardens and, interestingly, includes a simplified sketch of a style of gate that appeared in a photograph from Cleeve Prior Manor, some 12 miles from Hidcote, in *Gardens for Small Country Houses* by Gertrude Jekyll & Lawrence Weaver. The Cleeve Prior gate is similar to those used in several places at Hidcote. Other books borrowed on garden design were H. Inigo Triggs's book *Garden Craft in Europe* (1913) which is a history of garden design that shows how great garden designers have used landscaping, plants, ornaments and water to create a style for every time and place; John D. Sedding's book *Garden-craft old and new* (1903); and Gertrude Jekyll's book with paintings by George S. Elgood entitled *Some English Gardens* (1904) describing thirty-six British gardens attached to great houses or castles. These indicate that Lawrence was thinking ahead to when he would be back at Hidcote and able to continue work on creating the garden. The book by Inigo Triggs included a photograph of a similar feature in the garden at Versailles to that which Lawrence created in the Stilt Garden at Hidcote.

Other books that he borrowed then suggests that Lawrence was thinking about plant shelters and greenhouses for Hidcote as he borrowed K.L. Davidson's *The Unheated Greenhouse* (1907), Paul N. Hasluck's *Greenhouse and Conservatory Construction and Heating* (1907), W. J. May's *Greenhouse Management for Amateurs: the best greenhouses and frames, with details of their construction and heating; descriptions of the most suitable plants, with general and specific cultural directions; and all necessary information for the guidance of the amateur* (1911) and William F. Rowles's *The Garden Under Glass* (1914).

Lawrence's Army Records show that a further Medical Board that assembled at the 2nd Southern General Hospital, Bristol on 13 May 1915 examined him and recorded that *he has quite recovered.* His address was noted as being *Hidcote Manor, Campden, Glos.* He remained in England for a further year before returning to serve in France on 11 June 1916.

During this year at Hidcote Manor, it is likely that Lawrence planted the hornbeams in the Stilt Garden to the west of the Gazebos thus continuing the axis running from under the Cedar of Lebanon through the Old Garden, the Circle and the Red Borders. The photograph below shows the newly planted hornbeam stilts.

Red Borders looking towards newly planted Stilt Garden ca. 1916

This photograph shows that the urns were in place on either side on the top of the steps up to the Gazebos and that there were two large trees behind the northern Gazebo. An enlargement of the part of the photograph with the newly planted hornbeam stilts shows that the walls beyond which frame 'Heaven's Gate' had been built. However, the pillars on either side did not have the cherubs that now top them.

The Gazebos and the stilts and Heaven's Gate beyond ca 1916

Lawrence then returned to France to rejoin the Northumberland Hussars which had in May 1916 been reformed as a regiment at Briquemesnil on the Somme and attached to the XIII Corps for duty as Corps Cavalry. On 15 June, Lt. Col. Cookson took command of the regiment with Major Johnston as second in command. The regiment then received orders to proceed to Vaux-sur-Somme prior to the start of the great Somme offensive on 1 July. Early the following year, on 7 February 1917 Col. Cookson was transferred to the UK and Major Johnston was appointed the acting Commanding Officer of the regiment until command was assumed on 21 February by Major Reynolds.

Interestingly, Lawrence's mother, Gertrude Winthrop, applied to the American Embassy in London on 19 March 1917 for a passport for use in visiting the *United Kingdom* for *residence* and to visit *France* in order to *visit son.* The application says that she does not know when she will return to the United States but adds that she will do so *certainly after war.* It is not known whether Gertrude did indeed visit Lawrence in France.

In August 1917, Major Lawrence Johnston was selected for further promotion as he first attended the 1st. Army Company Commanders School at Hardelot, close to the French coast just south of Boulogne. Then in October/December 1917 he attended the Senior Officers Course at Aldershot. He was, however, unwell and on 14 December 1917, Brigadier Marshall, the Commandant of the Senior Officers School wrote to the General Headquarters, Home Forces, Horse Guards to say that *Major L. Johnston, 1/1st* [the first line regiment] *Northumberland Hussars, ... is I consider unfit to return overseas owing to ill-health, and I think in his case he should appear before a Medical Board.* His records do not show whether he did appear before a Medical Board. They do show that by April 1918 he was employed on liaison duties whilst serving with the British Mission to the French Army – for which he would have been well qualified as his record shows his special qualification was *French, fluent conversational.* Then from July through to September 1918 he served in the Reinforcement Camp of the Cavalry Corps.

Lawrence was clearly not well and returned on 7 September from Le Havre to Southampton on three weeks sick leave with his address being shown as *Boodles Club, St. James Street, London.* This club at 28 St. James Street is the second oldest gentleman's club in London with a distinguished membership over the years including nobles, politicians, historians, philosophers, writers and actors. A Medical Board on 16 December 1918 recommended a further two months leave for Lawrence until 16 February 1919. A note in his army file written by Lawrence requests permission *to proceed to the south of France for my health on Jan 2nd. 1919. My destination will be one of Lady Dudley's houses.* This refers to the hospitals for officers created by Rachel, Countess of Dudley, during World War I in France. These were at St. Nazaire, at Wimereux and in Paris. She also established the British Officers' Clubs that were first at Boulogne and then more widely throughout France that provided rest houses and a social rendezvous for hundreds of young officers.

Three months later Lawrence was back in England as he was demobilized, or as his records state *disembodied,* on 11 April 1919 when he was issued with a *Protection Certificate* by the Dispersal Hospital at Millbank, London, S.W.4. This declares that from that date *he will not be entitled to draw pay.* His medical category is C1 showing that he was fit for service in garrisons in the UK and could march 5 miles, see to shoot with glasses and hear well. It records his address as being Hidcote Manor, Campden and that his occupation in civil life was *Gentleman.*

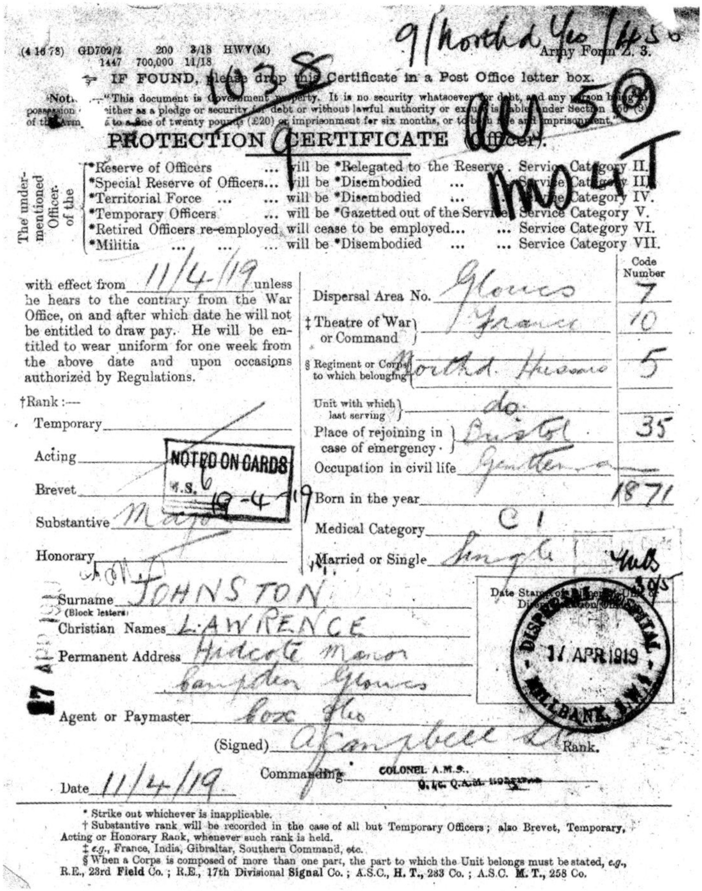

(4 1678) GD702/2 200 3/18 HWV(M)
1447 700,000 11/18
Army Form Z. 3.
9/North'd Yeo/1450
IF FOUND, please drop this Certificate in a Post Office letter box.

Note.—"This document is Government property. It is no security whatsoever for debt, and any person having possession of it either as a pledge or security for debt or without lawful authority or excuse is liable under Section ... of the Army ... to a fine of twenty pounds (£20) or imprisonment for six months, or to both such fine and imprisonment."

PROTECTION CERTIFICATE (Officer).

The undermentioned Officer of the
*Reserve of Officers ... will be *Relegated to the Reserve . Service Category II.
*Special Reserve of Officers... will be *Disembodied ... ... Service Category III.
*Territorial Force ... ... will be *Disembodied ... ... Service Category IV.
*Temporary Officers ... will be *Gazetted out of the Service Service Category V.
*Retired Officers re-employed will cease to be employed... ... Service Category VI.
*Militia ... ... will be *Disembodied ... ... Service Category VII.

with effect from 11/4/19 unless he hears to the contrary from the War Office, on and after which date he will not be entitled to draw pay. He will be entitled to wear uniform for one week from the above date and upon occasions authorized by Regulations.

		Code Number
Dispersal Area No.	Gloucs	7
‡Theatre of War or Command	France	10
§ Regiment or Corps to which belonging	North'd Hussars	5
Unit with which last serving	do.	
Place of rejoining in case of emergency	Bristol	35
Occupation in civil life	Gentleman	
Born in the year		1871
Medical Category	C 1	
Married or Single	Single	

†Rank:—
Temporary
Acting
Brevet
Substantive Major
Honorary

NOTED ON CARDS
19-4-19

Surname (Block letters) JOHNSTON
Christian Names LAWRENCE
Permanent Address Hidcote Manor
Campden Gloucs
Agent or Paymaster Cox & Co

Date Stamp of Dispersal Unit

DISPERSAL HOSPITAL
17 APR 1919
MILLBANK, S.W.1

(Signed) A Campbell Lt Rank.
Commanding COLONEL A.M.S., O.i/c Q.A.M. Hospital
Date 11/4/19

* Strike out whichever is inapplicable.
† Substantive rank will be recorded in the case of all but Temporary Officers; also Brevet, Temporary, Acting or Honorary Rank, whenever such rank is held.
‡ e.g., France, India, Gibraltar, Southern Command, etc.
§ When a Corps is composed of more than one part, the part to which the Unit belongs must be stated, e.g., R.E., 23rd Field Co.; R.E., 17th Divisional Signal Co.; A.S.C., H. T., 283 Co.; A.S.C. M. T., 258 Co.

Protection Certificate issued to Lawrence Johnston, 11 April 1919

Lawrence was now back at Hidcote and able to start thinking about resuming work on creating the garden. Soon after he returned, his mother bought the farm house and cottages as well as land adjacent to the garden of Hidcote Manor for the sum of £3,000 from Thomas Baldwyn, the owner of Hill Farm. This purchase was completed on the quarter day, 29 September 1919. The extract from the map showing the purchase makes it clear that the land transferred to Gertrude is that listed in the Conveyance as including *80 The Orchard, 58 Building and Orchard, 59 Two Cottages and Orchard* and *86 The Piece.*

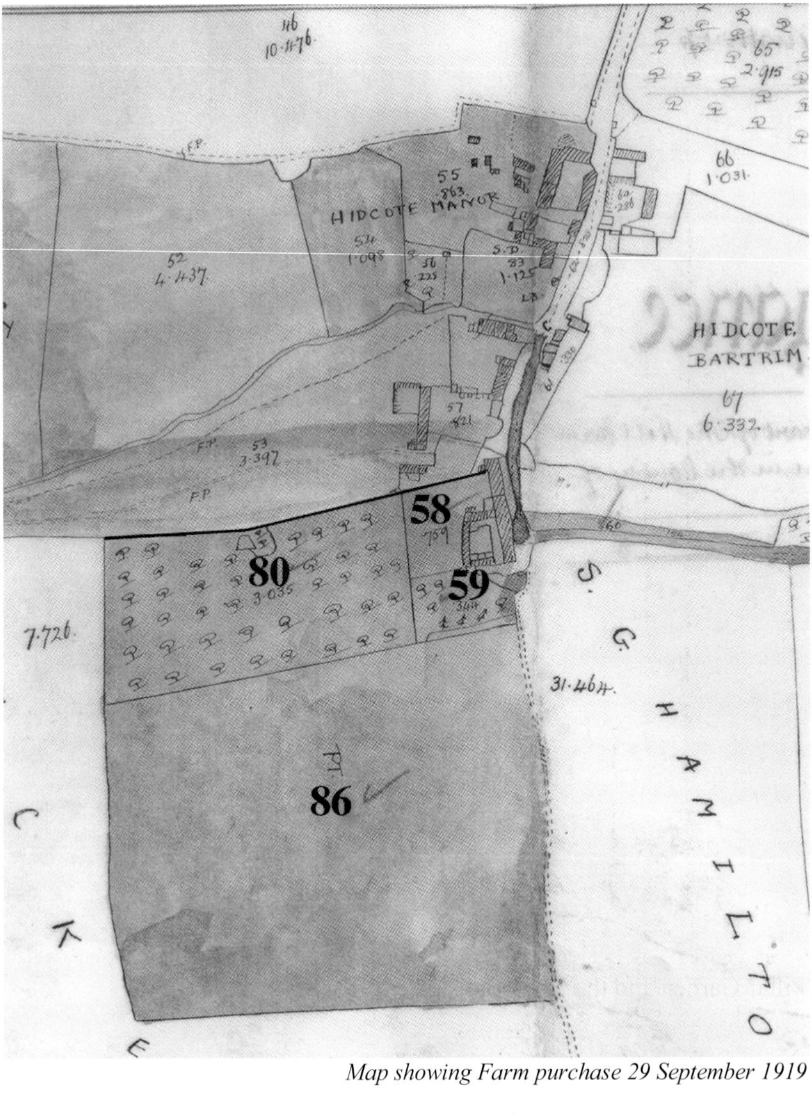

Map showing Farm purchase 29 September 1919

This purchase enabled the garden at Hidcote to be extended to the south through the orchard marked as 80 to the boundary fence with the field marked 86 – which remains the boundary of the garden to this day. The scene was therefore set in 1919 for Lawrence to continue to create the garden.

SIX

HIDCOTE MANOR AND GARDEN 1919 – LATE 1920s

Now that Lawrence Johnston had returned from serving with the Northumberland Hussars in World War I and his mother had purchased the land to the south of the garden, he was able to continue to create the garden at Hidcote. This next phase was to see the extension of the Long Walk and the creation of the Wilderness running southward out to the new boundary of the garden as well as the creation of new areas – Mrs. Winthrop's Garden, the Pillar Garden and the Rock Bank – within the area of the garden lying to the south of the axis from the Cedar of Lebanon westwards to Heaven's Gate and on the northern side of the stream running through the garden from east to west. The Theatre Lawn was also completed at this time with its yew hedges and to the north the creation of a Plant House for tender plants. It was also a time when some of the existing features were refined – such as the changing of the gateway from the Old Garden into the Circle and the raising of the pool in the Bathing Pool Garden.

The 1944 plan of Hidcote on the next page is annotated with these new areas running to the south – the Wilderness and the Long Walk – as well as with the new areas within the garden between the central east-west axis and the stream – the Lime Arbour, Mrs. Winthrop's Garden, the Pillar Garden and the Rock Bank.

The Long Walk was created as an axis running to the south from the southern Gazebo at right angles to the main central axis of the garden from the Cedar of Lebanon to Heaven's Gate. It was clearly planted in phases with the first phase running from the Gazebo down to the stream and then the second phase running from the stream out to the gates that form the termination of the Long Walk adjacent to the southern boundary of the garden. Lawrence demonstrated his understanding of perspective as he planted the Long Walk so that its width was wider (11 metre 25 cm) close to the stream and narrower (10 metre 30 cm) close to the gates at the southern end; a convergence of close to a metre.

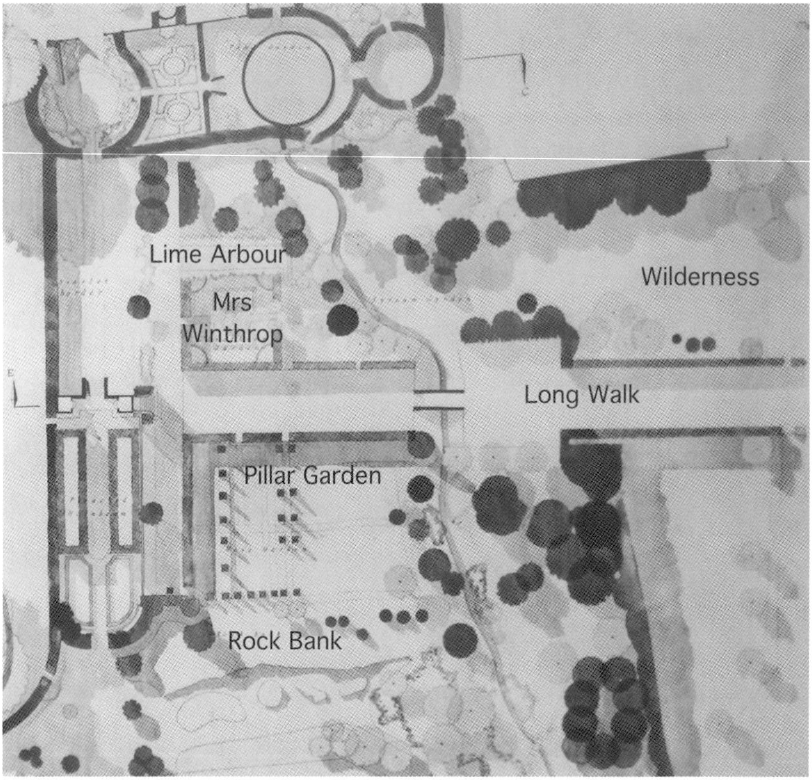

Annotated 1944 Hidcote Survey plan

An early photograph shows two large trees on either side of the bridge over the stream and whilst the Stilt Garden appears to be mature, there is no hedge between the Stilt Garden and the Theatre Lawn although a trellis work is in place. The steps down from the southern Gazebo have a circular pot on the bottom step and an adjacent rectangular pot on the ground on both sides. As with the earlier Maple Garden Lawrence was complementing the features of the garden with pots of plants. There is no sign of the shelter over the Alpine Terrace which was later to be constructed running to the west from the western side of the steps down from the Gazebo.

Long Walk towards Gazebo ca 1920

In a later view looking south from the Gazebo the two large trees on either side of the bridge across the stream have been removed and the shadows of the hedge beyond the stream show that this has only recently been planted.

Long Walk southwards from the Gazebo ca 1930

In designing the Long Walk, Lawrence was again following the principles set out by Thomas Mawson in *The Art and Craft of Garden Making* where he said that *Nothing is prettier than a vista through the smooth-shaven green alley or an archway framing a view of the country beyond.* And again, according to Mawson *Gates may advantageously mark the end of a vista.*

A view looking through the gates up the Long Walk show the detail of the pillars and the wrought iron gates.

Long Walk looking northwards ca 1930

The other development to the south of the stream was the creation of the Wilderness. In this, Lawrence was seeking a wild garden – a topic on which Mawson pointed out that *the infinity of curve in tree forms, the silent complexity of their myriad parts, their natural architecture which suggests no plan, alike baffle and defy description or dissection.* Mawson went on to say that there was, however, a leading principle *Unity in composition – to dispose all the trees, single, groves and clumps, to fill a place in one complete, unified scheme, and even to incorporate therein other pleasing foliage effects beyond.* In creating the Wilderness, Lawrence would have known of arboretums at other gardens such as the one at Westonbirt to which he sent plants from Hidcote in 1929 and also the one at Batsford, much closer to Hidcote than that at Westonbirt.

Looking south through the Wilderness, 2009

Within the Wilderness, Lawrence created a pond for flamingos. It is probable that he also had other species as there are photographs of ostriches and his diary in 1932 included the following list of ornamental birds which he may have been considering getting:

Kagus
Curassows
Scarlet Flamingos
" Ibis
Javanese Peafowl
Military Macaw ?
Argus pheasants ?

One view of the pond in the Wilderness shows three flamingos:

Wilderness, Flamingo Pool ca 1935

Lawrence also created a series of garden rooms on the south-facing slope between the central axis from the Cedar of Lebanon to Heaven's Gate. One was Mrs. Winthrop's Garden bounded on the north by beech hedging, on the west by hornbeam hedging and on the east by the limes of the Lime Arbour but open to the south with a view through The Wilderness. The photographs in the early 1920s show that the slender sundial, which had originally been in the White Garden, is now centrally located in the brick paved circle in the middle of Mrs. Winthrop's Garden.

This slender sundial is made up of a pilaster taken from the second bridge across the River Thames at Richmond as the square top of the pilaster has the following cut into the four edges 'Kew Bridge', 'Payne', 1789' and '1899'. The second Kew Bridge was built by Thomas Payne (or Paine) and opened on 22 September 1789 by King George III driving over leading *a great concourse of carriages.* The date of 1899 corresponds with the date when the second bridge was demolished in October to December 1899.

Mrs. Winthrop's Garden looking north

To the east of Mrs. Winthrop's Garden is the Lime Arbour.

Looking south through the Lime Arbour 2009

In the Lime Arbour Lawrence cleverly enhanced the perspective by planting the limes so that the distance between the boles on the two sides are 13 feet 6 inches (4.1 metres) apart at the northern end and 10 feet 6 inches (3.2 metres) at the southern end – a convergence of some three feet (0.9 metres).

To the west of the Long Walk, the Pillar Garden was created as a Mediterranean garden, with rows of yew pillars on a series of shallow terraces, enclosed by a beech hedge to the north and a hornbeam hedge to the east bordering the Long Walk. A photograph from 1930 looking towards the northern beech hedge and the Stilt Garden beyond shows the slender pillars of that day together with the rows of paeonies – many of which are still in the Pillar Garden today.

Pillar Garden looking north ca 1930

The pillars today in a view looking westwards are much more sizeable but still rise as in the 1930s from a cube at their bases.

Pillar Garden looking westwards 2007

To the west of the Pillar Garden, Lawrence created the Rock Bank to simulate going from the Mediterranean up to the Alpes-Maritimes, thus following Mawson's advice that rock gardens *ought to have, above all things else, a definite plan, and should aim to reproduce some particular phase of nature.* He also followed the recommendation that it be *secluded from the purely ornamental and formal parts of the grounds by a screen of pines and dark foliaged evergreens.* The choice of simulating the Alpes-Maritimes probably reflected the fact that Lawrence and his mother in the early 1920s visited Menton on the French Mediterranean coast close to the border with Italy and from 1924 Lawrence began to purchase land in the Val de Gorbio inland from Menton to create a garden at Serre de la Madone. In the summer months Lawrence would travel north into the Alpes-Maritimes to escape the heat of the Mediterranean coast by staying in a rented house in the Vallée des Merveilles that is now part of the Mercantour National Park. Consequently, Lawrence was well aware of the terrain that he was simulating in the Rock Bank at Hidcote.

Rock Bank ca 1930

Into the Rock Bank, Lawrence incorporated a feature to simulate the glacial melt – this has recently been rediscovered and restored. This again echoes Mawson's maxim that *Unless water accompanies the rock garden in some form or another, it never seems complete.*

Melt pool in Rock Bank 2010

The Great Lawn – now known as the Theatre Lawn – with its yew hedges and beech tree on the raised dais at the western end was completed during these years. Old photographs from the early 1930s show that the Great Lawn was being used to play bowls.

Theatre Lawn ca 1960s

To the north of the Great Lawn, a plant house was erected to provide shelter for tender plants. This had glass panels on the front which could be removed for the summer months. In creating this, Lawrence will have had Mawson's advice in mind that *No garden feature is to be considered as a thing apart, and should fall inevitably into its place, being designed and detailed in accord with the surroundings.* It is evident that the plant house at Hidcote was intended as being more of a conservatory in which Lawrence and his guests could sit and enjoy the garden at any time of the year. As Mawson said *There is perhaps no detail of domestic architecture which calls for so much care in its design and proportions as a conservatory. ... it is a feature which may be a delight to the eye as well as forming a most useful and pleasant adjunct to the mansion.* He went on to add that *In most circumstances a pleasant "withdrawing" room is required where, at all periods of the year, the users may enjoy the sunshine amid fresh flowers and foliage...*

Plant House ca 1930

Further shelter was provided for the Alpine Terrace by roof panels that sheltered the alpine plants from the winter rain of the Cotswolds – these panels were removed during the summer months – and by a conservatory which today is the extension of the restaurant.

These years also saw some changes to the earlier garden structure. Thus the Bathing Pool Garden that initially featured a ground level pool was enlarged and raised by surrounding it by a low wall some 0.5 metre (18 inches) above the surrounding ground, increasing the depth to 0.64 metre (2 feet 1 inch) deep. In doing so Lawrence was following Mawson's suggestions of a plain circular pool with a wall 18 inches high and a depth of 2 feet 3 inches almost to the letter and furthermore that:

... The best place for a fountain is an enclosed court of some kind ... In such cases the light feathery streams may rise from the surface of

the water, or where more elaboration is called for, a group of statuary, such as the boy with a dolphin may be introduced.

Hidcote fountain

Mawson boy with dolphin

Bathing Pool Garden ca 1930

A further change was to the gate between the Old Garden and the Circle that initially had a fairly narrow opening:

Gate from Old Garden to The Circle ca 1910

This was changed to a much wider gate with the gate posts being moved to the sides:

Gate from Old Garden to the Circle 2009

The years following World War I saw Lawrence becoming good friends with his contemporaries – the landowners who put their energy into creating gardens. He was already a close friend of Mark Fenwick who came in 1901 to Abbotswood on the western slope of the hill up to Stow-on-the-Wold. Fenwick came from a wealthy Northumberland family and had the house at Abbotswood extended and remodelled by Edwin Lutyens who also designed the gardens to the south and west of the house. Mark Fenwick was described by Lord Redesdale who created Batsford Park, another garden within 10 miles of Hidcote, as being *by far the best all-round amateur gardener that I know. His knowledge of his plants and their possibilities is really consummate. . . . He has worked at Abbotswood in such a way as to combine the formality of an Italian architectural garden with the broader and wilder lines of the natural woodland scene, the one fading into the other by the skill of imperceptible gradations.*

March 1920 was to see the idea being put forward of a Garden Society. This was first proposed by Mr. J.E. Williams of Hillside, Llandaff in a discussion in March 1920 with Mr. W. R. Dykes, the recently appointed Secretary of the Royal Horticultural Society. Following further discussions in May 1920 between Mr Dykes, Mr. Charles Nix (of Tilgate Forest Lodge, Crawley), Mr Reginald Cory (of Dyffryn, Vale of Glamorgan) and Mr Gerald Loder (of Wakehurst Place Garden, West Sussex) it was decided to ask a certain number of amateur gardeners to dine together on the evening of the first day of the Chelsea Show with a view to forming a Society. A circular signed by Reginald Cory, Gerald Loder and Charles Nix was accordingly sent out to a number of people inviting them to dine informally on 1st June 1920 at Jules Restaurant, Jermyn Street at a cost of 12s 6d a head. The intention was that those should be invited *who not only possessed gardens but were actively engaged in the cultivation, increase and exchange of plants, and especially plants of more recent introduction.*

It was agreed at a dinner in November 1920 that the club should be called *The Garden Society* and that the number of members should not exceed 40 and all members must be Fellows of the Royal Horticultural Society and be male. The Garden Society would meet at least twice a year and dine together on the first night of the RHS Spring Meeting now held at Chelsea and on the night of the first RHS meeting in November. At the next dinner in April 1921 when Mark Fenwick was elected a member it was agreed to increase the number of members to 50. Later the same year, in October, Mark Fenwick wrote to Lionel de Rothschild (of Exbury, Hampshire) to ask him if he would support his, Mark Fenwick's, nomination of Lawrence Johnston *who is a very keen & very good gardener.*

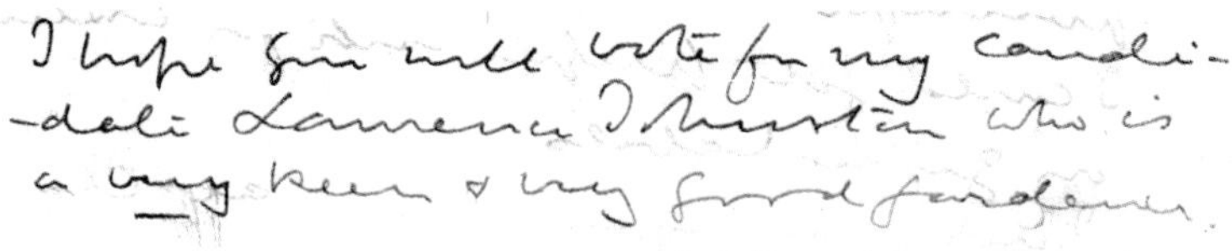
I hope you will vote for my candi-
-date Lawrence Johnston who is
a very keen & very good gardener.

Mark Fenwick letter, 10 October 1921

At the seventh dinner on 23 May 1922, Lawrence Johnston and the Marquis of Headfort (of Kells, County Meath, Ireland) were both elected members. Interestingly, even though Lawrence would not have been present, this was the dinner at which Ernest H. Wilson, who was the plant hunter born in Chipping Campden, as a guest gave an interesting account of his journeys and work in introducing plants from China and elsewhere.

The motto of the Garden Society – *Petimus Damusque Vicissim* – was selected by Gerald Loder from Horace's Ars Poetica and translated it means *Turn by turn we ask and give.* This reflected the way that the Garden Society enabled its members to bring flowers of interest and to make a few remarks about them leading through the personal exchange of views and information to the exchange of plants.

Lawrence will have come to know all the members of the Garden Society through the requirement to attend dinners at least once every two years – as the Society decided that any member who failed to attend a dinner of the Society for two years should automatically cease to be a member. He will thus have known all of his contemporaries who were enthusiastic landowners who put their energies into creating gardens.

The 1920s also saw the appointment of Lawrence's first professional head gardener at Hidcote – Frank J. Adams – who came to Hidcote in 1922 from Windsor Castle where he had been in charge of flower decorations. Adams was to stay at Hidcote until his death in 1939.

Frank J. Adams (on left) with Lawrence in Mrs. Winthrop's Garden, ca 1925

An earlier photograph of Lawrence with his gardeners at Hidcote dates from the 1910s.

Lawrence Johnston and his gardeners, early 1910s

In the years after the Great War, the Farm bailiff at Hidcote was George Lennox Wheeler who continued in that post up to the time of World War II. The farm raised sheep during these and later years as Lawrence received awards for them in 1928 and 1934 at the annual Campden Teg [2 year old sheep] Show and Sale.

An interesting insight into how the Manor House looked and was furnished in these post war years can be gained from an inventory of household furniture carried out by Bosley & Harper of Shipston-on-Stour in 1921. The rooms listed are as follows: *Hall, Major Johnston's Study, Inner Hall by Drawing Room, Drawing Room, Staircase* and then upstairs are *Bedroom No. 1, Bathroom, Bedroom No. 2, Bedroom No. 3 (Major Johnston's)* and *Vestibule to Bedroom No. 3*. These rooms correspond to the rooms shown on the 1913 plan of the Manor House.

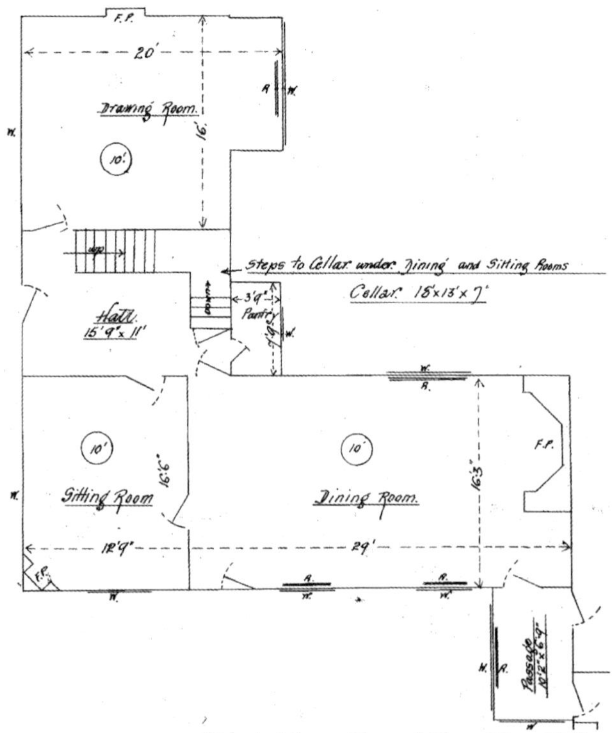

Hidcote Manor Ground Floor Plan, 1913

The Hall in the inventory corresponds to the *Dining Room* in the 1913 plan, *Major Johnston's Study* to the *Sitting Room* and the *Inner Hall by Drawing Room* to the *Hall.*

On the first floor, it is evident that Bedroom No. 1 in the inventory is the one 20 feet long by 16 feet 2 inches wide, Bedroom No. 2 the one 16 feet by 13 feet, the Bathroom as shown and Bedroom No. 3 the one 16 feet 4 inches by 16 feet with the adjoining vestibule shown as a Passage.

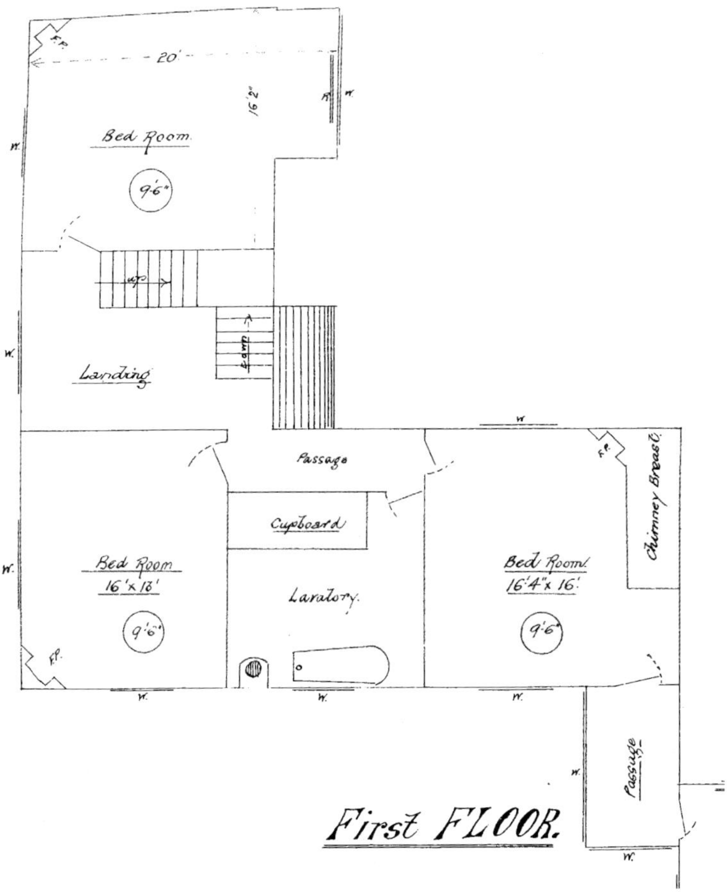

Hidcote Manor, First Floor Plan, 1913

There is no mention in this inventory of any rooms on the second floor of the Manor House or in the extension – presumably these were the servant's quarters and thus did not contain items of furniture that merited listing in the inventory.

The inventory of 1921 shows that the Hall was furnished with a *carved oak refectory table with carved bulbous legs* and an oak dresser and oak settle together with eight oak chairs with leather-backs. There was also a high-backed easy chair with a green tapestry cover which probably was near the fireplace. On the wall opposite the door was an 8 foot 9" high 11 foot 3" long tapestry of the Sermon on the Mount. Oil paintings on the other walls included two of horses and one called *Hunters and Hounds.*

Major Johnston's study contained a *walnut escritoire with an enclosed bookcase over* together with an *oak writing table.* There were several chairs as well as two lounge chairs and a pitch pine settee with a chintz cover. On the walls were two engravings of *The Rustic Hovel* and *The Cottage Sty* as well as two engravings after George Stubbs entitled *Shooting.*

The Inner Hall had a gothic oak chest, a double oak corner cupboard, some chairs and oak table with a mahogany barometer and thermometer on the wall together with an oval mirror and a painting of a Hereford Ox and four sporting prints.

On the Staircase there were more engravings of *Woodcock Shooting* and *The Return,* as well as one of *Oliver Cromwell.* There were also, presumably on the landing, a *Queen Anne walnut escritoire,* an *8 day Grandfather clock* in a walnut case and two three-tier bookcases. Further engravings were of *Going out in the Morning* and *The Death.*

The Drawing Room contained a lacquer cabinet with three drawers under, a walnut writing table and a mahogany card table. There were two William and Mary chairs, a mahogany Chesterfield couch in cretonne with a chintz cover and various other chairs. There was a grandfather clock and, on the walls, a William and Mary bevel plate glass mirror in a gilt frame, an oil painting of *Fruit and Flowers* and miniature portraits of ladies.

Bedroom No. 1 contained an oak four post bedstead with rare old French linen bed furniture whilst Major Johnston's bedroom (No. 3) had an oak bedstead with carved panel head rail together with a set of 18th Century bed hangings in a plum colour. His bedroom also had a painting entitled *Country Scene*. All three bedrooms had a dressing table as well as a wash stand and its associated china toiletware. The vestibule to his bedroom had a revolving bookcase and a couple of chairs.

These descriptions enable us to envisage how the Manor House must have looked in the early 1920s.

These post war years saw Gertrude Winthrop being relatively settled at Hidcote. Following the war, Mrs. Winthrop had built a cover over the village well and had marked this with a carved date, commemorating the end of the war, as a memorial to those who had died.

Great War memorial, Hidcote Bartrim ca 1920

The engraving on the stone on the war memorial marks the date of the end of World War I as it bears the legend Nov. 11 1918 and is also engraved with G W. to signify the Great War.

Great War memorial, Hidcote Bartrim

The building of this memorial by Gertrude is in keeping with the care that she took over the families which lived in the village. For the spiritual improvement of the villagers, she had a cottage converted into a chapel and arranged for lay preachers to take Sunday afternoon services there. It is said that Gertrude was soon after anyone who did not attend. On one occasion on 4 November 1921 she organised a combined Whist Drive and Dance in King George's Hall, Mickleton to raise funds for the Red Cross. It was reported that *36 tables sat down to the whist and attendance at the dance was far too large for comfort. Mrs. Winthrop generously provided both the prizes and the refreshments.* The article concluded by saying that *Mrs. Winthrop was applauded by the entire company at the end.*

A photograph taken in the early 1920s shows Gertrude in the garden at Hidcote whilst another, in the same setting, shows her butler, William Brown, who had come to Hidcote in about 1910 and died the year after Gertrude in 1927.

Gertrude Winthrop 1920s

William Brown, butler, 1920s

These photographs, taken on the Cedar Lawn at Hidcote with the Manor House behind, were probably taken on the same date as another photograph of Gertrude with her staff at Hidcote as they have the same items of garden furniture and Gertrude is wearing the same outfit. The men who are shown with Gertrude are likely to have included those shown in the electoral roll as living in the village in the early 1920s: Frank Adams, George, Leonard and Walter Bennett, George and Alfred Holtom, William and John Hughes, Robert Payne, Edward Pearce and George Wheeler (who was the farm bailiff). Other Hidcote staff such as William Brown, the butler, lived in Mickleton and Ernest Daniels, the chauffeur, lived in Ebrington.

Gertrude Winthrop and her staff at Hidcote, 1920s.

From Gertrude's passport applications and the records of immigration into the United States and the United Kingdom, after coming to Hidcote in 1907, Gertrude made three visits to the United States – the first in July 1907 when she was accompanied by Miss Hedwig von Lekow (returning in June 1908), the second in May 1912 when she was accompanied by Lawrence and Miss Hedwig von Lekow (returning in July 1912) and the third in May 1921 when she was accompanied by Ellen Graham (returning in August 1921).

Her passport application on 21 October 1920 to the American Consul-General in London says that she had last visited the United States in 1912 and wished to have a passport in order to visit France for her health and to reside in the British Isles. This application declares her height to be 5 feet 2 inches and that she had blue eyes, white hair and a fair complexion. A further application submitted on 4 September 1922 states that her last visit to the USA was in 1921. On this occasion, however, the reason for seeking a passport is much less informative – simply indicating travel to all countries. Attached to this application is an *Affidavit to Explain Protracted Foreign Residence and to Overcome Presumption of Expatriation* in which Gertrude declares that she ceased to reside in the United States in June 1908 and has subsequently resided temporarily in England at Campden, Glos., Eng. and that the reason for such residence being *To keep house for my only son.* The affidavit goes on to set out the ties of family, business and property with the United States as being *Brothers, sister, nieces and nephews residing in the U.S.A., Money invested in America and I own a house at 40 East 69th Street, New York City.* Gertrude also confirms that she pays American income tax at New York. Interestingly, Gertrude's application for her passport in 1922 was witnessed by Ellen Graham who declares that she lived in London and Campden, Glos. and has known Gertrude for the past 8 years. This shows that she had known Gertrude since 1914 and it is probable that the reference to knowing her in London might result from Gertrude having lived in London – perhaps at the Hans Crescent Hotel where she had stayed in March 1913 and whose manager had witnessed a codicil in 1924 to Gertrude's will. Another occasion could have been when Lawrence was a patient in the King Edward VII Hospital for Officers from November 1914 to March 1915 as Gertrude's butler William Brown was one of those who had signed out books for Lawrence from the Lindley Library. Although Ellen Graham is described as a maid when she travelled first class with Gertrude to New York, it seems likely that she may have been more of a companion – and consequently able to provide the required affirmation that Gertrude was an American citizen.

As usual, Gertrude's visit to the USA was recorded in the Social Notes of the *New York Times* which on 6 June 1921 recorded as its first item that *Mrs Charles F. Winthrop of Hidcote Manor, Campden, Gloucestershire, England, is due to arrive today on the Lapland, and will be at the Gotham. Mrs. Winthrop and her son by her first husband, Captain* [sic] *Lawrence Johnston, have made their home in England and this is her first visit here in years.* The Gotham Hotel was at 5th Avenue and 55th Street and was one of the luxury hotels of its day. The Ellis Island immigration records show Gertrude arrived on the *S.S. Lapland* accompanied by her maid, Ellen Graham, aged 50. She and her maid had travelled first class and were intending to stay in New York for two months. They both returned on the *S.S. Rotterdam*, again travelling first class, arriving at Plymouth on 7 August 1921.

Gertrude, then in her seventies and approaching eighty, didn't like the winters at Hidcote and took to travelling to the South of France for the winter months. She and Lawrence went to Menton, on the Mediterranean coast of France, close to the border to Italy. In the early 1920s they stayed at the Villa Mer et Monts, in the Val de Gorbio, a few kilometres inland from Menton.

Villa Mer et Monts, val de Gorbio, Menton

A photograph shows Gertrude, on the left, having a meal at the Villa Mer et Monts.

Mrs. Winthrop (on left), Villa Mer et Monts, 6 April 1925

The butler is Fredo Rebuffo, Gertrude's butler at Menton, and the lady on the right is described as being Gertrude's companion – and may well be Ethel Graham who travelled with Gertrude to New York in 1921.

The 1920s had seen Lawrence's return from the Great War and his resumption of the creation of the garden at Hidcote. He appointed his first Head Gardener, Frank Adams, and Lawrence was elected a member of the Garden Society which brought him into a group of wealthy landowners who were keen gardeners. His mother, Gertrude, was approaching 80 and she and Lawrence began to spend their winters on the Mediterranean Coast at Menton.

SEVEN

SERRE DE LA MADONE

In the early 1920s Gertrude Winthrop, approaching the age of 80, took to spending her winters on the Mediterranean coast of France at Menton, known as *la perle de la France* (The Pearl of France), on the French Riviera close to the border to Italy. At the end of the 19th century, Menton was popular with the English and Russian aristocracy who built many of the luxurious villas and palaces that still grace Menton today.

Gertrude and Lawrence stayed at the Villa Mer et Monts a few kilometres inland from Menton in the Val de Gorbio. They clearly liked this area immensely because on 24 January 1924 Lawrence bought the first of several plots of land that over the years were going to make up the Serre de la Madone – which is situated adjacent to the Villa Mer et Monts – from Monsieur Fortune Ambroise Gioan and his wife. On the same day, he bought a second plot of land from Monsieur Bathelemy Caisson. Later in 1924, on 17 March, he bought another plot of land from the Fouilleron family. Four years later, on 2 June 1928, he bought more land and the house 'Casa Rocca' from the Rocca family. Then in 1930, on 30 May, he bought another piece of land along with the Villa Serre de la Madone from Monsieur Philippe Maldini and his wife. Yet another purchase, on 16 May 1933, was from Madamoiselle Julie Milhau, and on 15 January and 16 January 1936 two more pieces of land were obtained. The final acquisitions were on 26 April 1939 from Mademoiselle Jeanne Marie Anrigo and on 24 January 1940 again from Monsieur Philippe Maldini and his wife.

These successive purchases are shown on the following map of the property at Serre de la Madone.

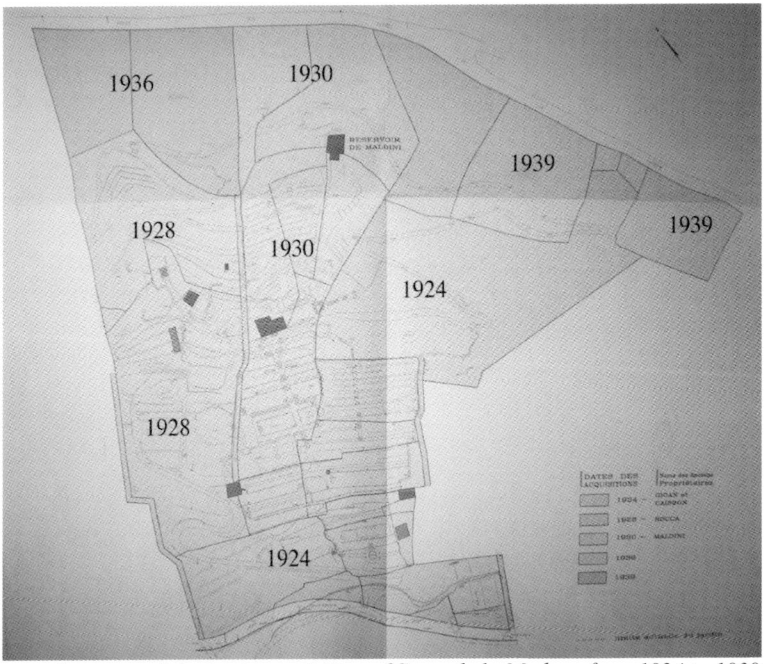

Lawrence Johnston's acquisition of Serre de la Madone from 1924 to 1939

Lawrence Johnston created his garden at Serre de la Madone from land originally planted with lemon trees, olive trees and vines on terraces running up the south-west facing steep slope of the Val de Gorbio.

↑ Villa Serre de la Madone ↑ Property boundary ↑ Villa Mer et Monts

The garden at Serre is made up of a series of terraces from its lowest level at 50 metres above sea-level to its highest level some 100 metres higher at the road running along the crest of the ridge. It is thus totally different from the much flatter land at Hidcote and has a quite different climate. It did mean, however, that any plants that Lawrence acquired could generally be grown successfully at one or other of his two gardens.

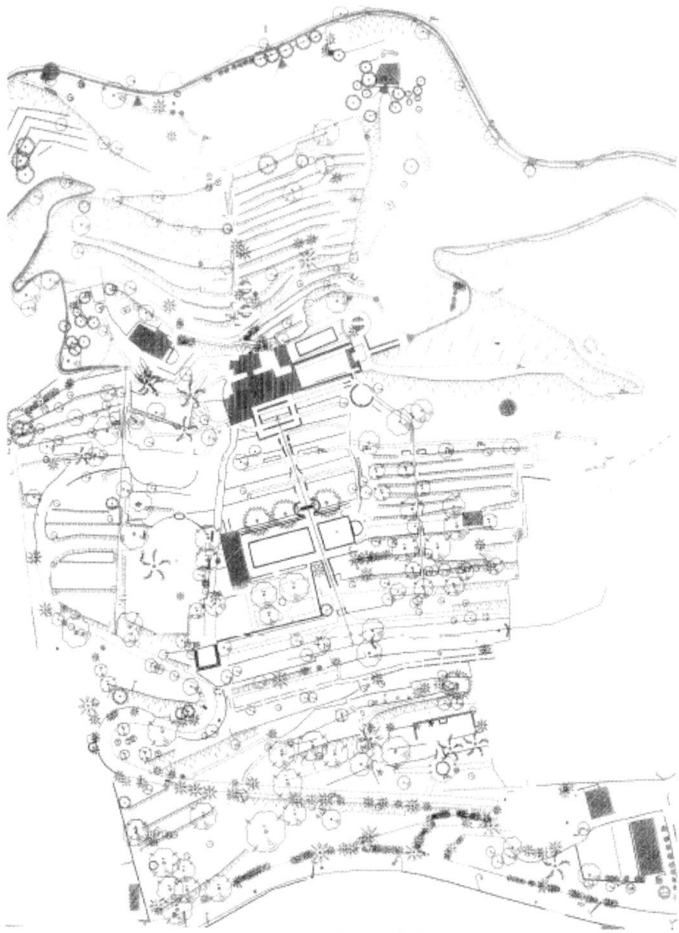

Serre de la Madone plan of the garden

At the lowest level, there was a garden with a cold greenhouse. At the next level, there was a shade garden with a long pergola, then a level with a

Mexican garden. Next was one with the square of four plane trees, and then the Winter Garden with a statue of Venus emerging from a pool of water.

The pool with a statue of Venus emerging from the water, 1937

From here, the Villa Serre de la Madone can be seen several terraces higher.

View looking up towards the Villa, 1937

The Plane Garden, 1937

The Bowling Green, 1937

As no photographs have yet emerged of the interior of Hidcote Manor, even though inventories have been found of the furniture and furnishings at Hidcote, it is of particular interest that there are several interior photographs of the villa at Serre de la Madone in 1937 as they show the style favoured by Lawrence.

The Sitting Room 1937

Looking the other way in the Sitting Room 1937

The Dining Room, 1937

Further rooms in the Villa 1937

Lawrence Johnston's bedroom 1937

A guest bedroom, 1937

Gertrude Winthrop's death

The early 1920s saw Gertrude becoming aware of her age and in early 1923 she approached the Vicar of St. Lawrence's Church in Mickleton about a mile down the hill from Hidcote, and which she had been attending regularly when she was at Hidcote, to seek *a legal facility to reserve a portion of the churchyard.* The minutes of a joint Vestry and Parochial Church Council meeting *called specially for the purpose* held on 1 February records:

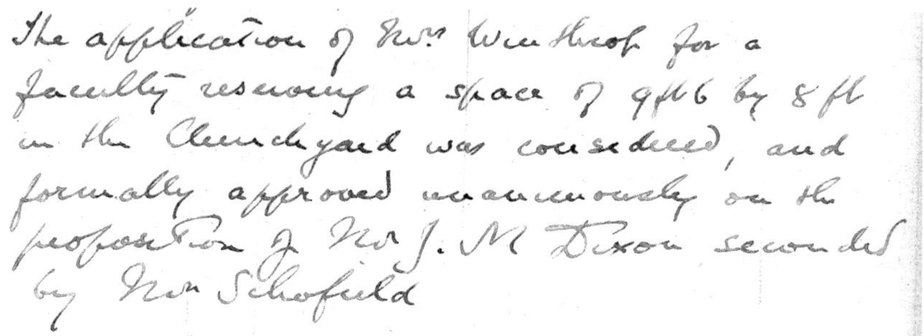
The application of Mrs Winthrop for a faculty reserving a space of 9ft 6 by 8ft in the Churchyard was considered, and formally approved unanimously on the proposition of Mr J. M Dixon seconded by Mr Schofield

St. Lawrence Church, Mickleton 1 February 1923

The following year, Gertrude, made two wills – one for her property in England and the other for her estate in the United States. These were both written on the same day, 7 November 1924, and both had the same witnesses: Valentine Worthington, Counsellor at Law, 34 Cheyne Court, Chelsea London SW 3, and Lucius F. Crane, Solicitor, Bush House, Aldwych London WC 2. Valentine Worthington was an American who was educated in England at Westminster School and then at Christ Church, Oxford and then practised law in New York City. Lucius F. Crane was a law graduate of the University of Cambridge and a Member of the New York Bar who published an article in a professional legal journal in 1925 which addressed the position of married women under American law. It is thus evident that Gertrude had taken the best professional advice in writing her wills so as to minimise any liability to death duties.

As Gertrude became older she began to lose her memory and in 1925 Lawrence Johnston, Edward Farquhar Buzzard (an outstanding doctor expert in neurology who was appointed physician-extraordinary to King George V in 1923), and Ellen Graham, Gertrude's maid and companion, swore affidavits that resulted in the Court of Protection in London issuing an authority on 11 August 1925 enabling Lawrence to receive on behalf of *the patient Gertrude Cleveland Winthrop* the rents and income and dividends paid to her. Johnston was also given authority to take steps to appoint the bank of New York & Trust Company as the sole administrator in America of the patient's property. Consequently, the New York Times on 14 May 1926 reported that *Supreme Court Justice Glennon yesterday appointed the Bank of New York and Trust Company* as administrator of the American property of Mrs. Gertrude Cleveland Winthrop. The article noted that the application was made by her son, Lawrence W. Johnston, *because a large part of his mother's property is in this country.* It adds that Gertrude has an income of $12,000 a year from her real estate and of $59,447 from her personal property (together corresponding to over $850,000 a year today).

Gertrude died at Villa Mer et Monts, Menton on 8 December 1926 just twelve days before her 81st birthday on 21 December. Her body was brought back to England and she was buried on Wednesday 15 December in the plot she had reserved in the churchyard of St. Lawrence in Mickleton, about a mile down the hill from Hidcote Bartrim. The report of the funeral in the *Evesham Advertiser* noted that the mourners included Major Johnston (son), Messrs. Gordon Woodhouse and William Barrington (of Lypiatt Manor, Stroud), the Rev. W. J. and Mrs. Guerrier (of Ebrington), Col. and Mrs. Gibbs (of Admington Court), Major and Mrs. Muir (of Kiftsgate Court) and Mr. and Mrs. Paul Woodroffe (of Campden). Those who sent wreaths included Violet Gordon Bell, Major and Mrs. Muir, Lady Saye and Sele of Broughton Castle, and Mr. and Mrs Bright-Smith. Gordon Woodhouse was the husband of Violet Woodhouse, a famous musician, whose household included William Barrington; they had initially lived at Armscote and were close friends of Lawrence. The Bright-Smiths were friends from Little Shelford and Vandeleur Bright Bright-Smith had been one of those who had signed Lawrence's application in 1900 to become a naturalized British citizen. The newspaper account also noted that *The bearers, who were employees of Mrs. Winthrop were as follows: Messrs. G. L. Wheeler, F. H. Adams, E. Pearce, W. Bennett, H. Holtom, W. Hughes sen, W. Hughes jun, A. Hawkins, and C. Lynes (of Shipston-on-Stour) in charge.* Mr. Wheeler was the farm manager at Hidcote, Frank Adams was the head gardener and Ted Pearce, Walter Bennett and Albert Hawkins were all gardeners.

Gertrude Winthrop grave, St. Lawrence Church, Mickleton

As already noted, Gertrude had made two wills. One will was *as to my property in the United Kingdom* whilst the second one was *as to my property in the United States.* The English will declares Gertrude's domicile as being in the State of New York in the United States although she is currently residing at Hidcote Manor, Chipping Campden, Gloucestershire. This will left bequests of £600 to her goddaughters, Ethel Boulton and Contessa Balbis of Castello di Sambuy near Turin, Italy as well as the same amount to Fraulein Hedvig von Lekow of Kitzbuhel, Austria. Various smaller bequests were left to her butler (William Brown) (£40), chauffeur (Ernest Daniels) (£60), maid (Ellen Graham) (£50), housemaid (May Bennett) (£25) and her first (Edward Pearce) (£25) and second gardeners (Edward Hughes) (£25). It leaves to Lawrence Johnston *all other my jewellery not hereinbefore specifically bequeathed and all my furniture books pictures silver and other household effects and all my personal effects and belongings of whatsoever nature at Hidcote Manor.* It also leaves to Lawrence:

Fifteenth. I give and bequeath to my son Lawrence Johnston all my cash in Lloyds Bank Stratford on Avon or in any other bank in England at the time of my death or in my possession in England at the time of my death and all moneys invested in England at the time of my death whether in War Loan or in other investments and I declare that this bequest shall also include all personal estate of mine situate in England at the time of my death

Gertrude Winthrop English will, 7 November 1924

as well as

Sixteenth. I give devise and bequeath to my son Lawrence Johnston all my property in England known as "Hidcote Manor" together with all the furniture and other contents thereof not hereinbefore specifically bequeathed and together with all the heredita-ments and appurtenances thereto belonging or in anywise appertaining

Gertrude Winthrop English will, 7 November 1924

Probate was granted on 11 February 1927 with the gross value of the estate in England being £15,853-6-9 (equivalent to about £750,000 today).

Gertrude's American will related to an estate, valued in the *New York Times* as being some $2 million (equivalent to $24 million today). She left two bequests – the first to her brother James M. Waterbury of an annuity of $10,000 a year and the second to her niece Katherine Livingston Waterbury, who had been an invalid since childhood, of an annuity of $3,000 – with the provision in both cases that on their death the capital shall form part of her residuary estate. She left to Lawrence *all the furniture and other contents of my house No. 40 East 69th Street New York City.* In addition, she left the income from her residuary estate to Lawrence with the provision that on his death the residuary estate shall pass to his children.

> Seventh: All the rest residue and remainder of my property and estate situate in the United States of America real personal or mixed, hereinafter called my residuary estate, I give devise and bequeath unto my trustees upon trust to invest and re-invest the same and to keep the same invested and to pay over the net income thereof to my son Lawrence Johnston during his life in equal quarterly payments. Upon the death of my said son I direct my trustees to distribute my residuary estate then held in trust as follows namely to convey transfer pay over and deliver the whole of my said residuary estate to the lawful issue of my said son if more than one in equal shares absolutely and forever.

Gertrude Winthrop American will, 7 November 1924

In the event, as Lawrence died in 1958 without having married and had children, the residuary estate under Gertrude's will then in 1958 passed to certain nephews and a niece of Gertrude in the Waterbury family. Although Gertrude left two annuities, both her brother James M. and her niece died in July 1931, thus increasing the annual income received by Lawrence from the residuary estate by $13,000 a year. As already mentioned, the annual income received just before her death by Gertrude from her estate in the United States amounted to about $72,000 (equivalent to about $850,000 a year today). Gertrude had taken good steps to minimise tax liabilities at a time when death duties in the UK were large.

So, in today's values, Lawrence was left an estate in England worth £¾ million and the income during his life of $850,000 a year from the estate in the United States. Gertrude had carefully taken steps to minimise death duties by arranging that after Lawrence's death the $24 million residual estate would be divided equally between his children – but he had none. Gertrude was clearly a loving mother who had arranged everything for her son's benefit.

EIGHT

PLANT HUNTING

It will be recalled that when Lawrence Johnston was recovering in the King Edward VII Hospital for Officers in London from being wounded at Ypres in October 1914 he was borrowing books from the RHS Lindley Library on plant hunting. These books included Frank Kingdon-Ward's *The Land of the Blue Poppy: Travels of a Naturalist in Eastern Tibet* which had been published in 1913. This described Kingdon-Ward's expedition from Burma, into the Hengduan Mountains of north-western Yunnan province, and then along the Mekong, Yangtze and Salween rivers in the region between eastern Tibet and western Sichuan. Another book that Johnston borrowed at this time was Ernest Wilson's *A Naturalist in Western China* – an account of eleven years travel and exploration in China – which had also been published in the previous year, 1913.

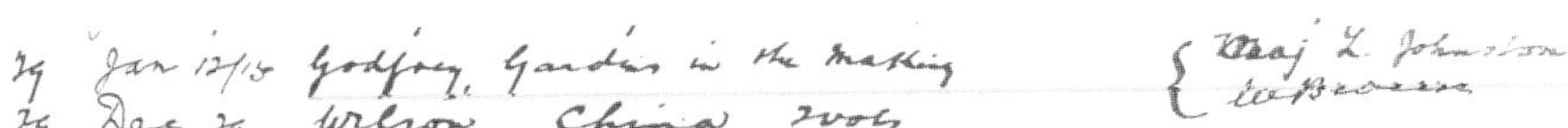
29 Jan 12/15 Godfrey, Gardens in the Making
29 Dec 29 Wilson, China 2vols
Maj L. Johnston

RHS Lindley Library loans, 1914/1915

This described Wilson's travels in China first on behalf of James Veitch & Sons and subsequently on behalf of the Arnold Arboretum in Boston. Although Ernest Wilson was born in 1876 in Chipping Campden, some 4 miles from Hidcote, he had been hired in 1906 by the Arnold Arboretum to collect seeds and plants from China. It is therefore unlikely that Lawrence, who arrived at Hidcote in the autumn of 1907 will have met Ernest Wilson.

In the 1920s after the Great War, Lawrence become involved in sponsoring and taking part in plant hunting as there was considerable competition among garden owners to be the first to have new spectacular plants. In the first decade or two of the twentieth century, plant hunters were sent out and funded primarily by commercial companies and nurseries such as James Veitch & Sons and Bees Ltd. The pattern and circumstances of sponsorship changed in the years after the Great War, with syndicates of wealthy garden owners beginning to club together to sponsor individual plant hunters. There were advantages over sole sponsorship as the costs and the risks and responsibilities were shared.

The Royal Botanic Gardens at Kew and Edinburgh frequently were part of the syndicates as they would make contributions in kind – by agreeing to receive the plants and seeds collected by the plant hunter and then distributing them to those who had purchased shares in the particular expedition. The shareholder received plants and seeds in proportion to the payment that he had made to the syndicate, or in the later years plants and seeds of the particular type of plant for which he had subscribed.

Over the next decade, Lawrence took part in plant hunting expeditions or sponsoring such expeditions to most of the continents – to Europe, to North and South America, to Africa and to Asia. As he had gardens at both Hidcote and at Serre de la Madone in the south of France, the seeds and plants that he obtained from these expeditions could go to one or other of his gardens.

The Swiss Alps, Lauteret Summer 1922

Lawrence went plant hunting in the Swiss Alps with E. A. Bowles, a fellow member of the Garden Society, in the summer of 1922 staying at Lauteret, near Grenoble. They will have visited the Lautaret Alpine Botanical Garden which was created in 1899 by the joint efforts of the University of Grenoble, the hotel manager at the Lauteret pass and the Touring Club of France. It was decided to create a garden here because it was one of the few places at high altitude (2000 m) that was accessible by car at the end of the 19th Century. Moreover, its geographical location where the northern and southern Alps met was one that favoured a great biological diversity with about 1500 species. Other members of the party included Dick (R.D.) Trotter (another Garden Society member) and Mrs Garnett-Botfield, who listed plants collected in her diary.

View of the Col du Lauteret

It was a short trip as they left on 29 June travelling by train and bus to the Hotel des Glaciers at Lautaret from which they made daily excursions. They were back in London by 11 July.

Syndicate for W. Th. Goethe, Andes, Autumn 1923

The following year, 1923, Lawrence became a member of a syndicate which sponsored a German plant hunter, W. Th. Goethe to visit the Andes to collect plants and seeds from the province of Neuquen in Argentina. Goethe had been collecting on Mount Whitney, the highest mountain in the continental United States, for A. K. Bulley of Bees Ltd. for two years. Bulley asked him if he would go plant-hunting in the Andes and, then, for reasons unknown, Bulley wrote to Henry D. McLaren at Bodnant asking him to take on the responsibility for the sponsoring of the expedition. There were six subscribers each paying £100 who, apart from Bulley, were all members of the Garden Society: A. K. Bulley, Lawrence Johnston, Sir William Lawrence, Lionel de Rothschild, Mr F. C. Stern and Henry McLaren. A letter of 9th July 1923 from Henry McLaren to Lionel de Rothschild shows the subscribers:

Mr A.K.Bulley,
Mr Lawrence Johnston,
Sir William Lawrence,
Yourself,
Mr Stern,
Myself.

Extract from 9 July 1923 letter from Henry D. McLaren to Lionel Rothschild

Goethe sailed from San Francisco on 20 July 1923 to Valparaiso, Chile and then travelled by train south in Chile to Puerto Varas when he travelled east across the Andes into Argentina to the town of Beriloche on the Lake Nahuel Huapi close to the source of the River Limay. From there he travelled to the city of Neuquen where he bought animals and tents to enable him to explore the foothills and mountains of the Andes. A letter on 10 September 1923 from Goethe to Henry McLaren makes it clear that his arrival at Beriloche had been delayed because of difficulties with the Chilean railroads as he says:

> *You are, no doubt, surprised that my arrival here is so late. I have had very considerable trouble in using the Chilean railroads, which I consider unfit to use for people coming from foreign countries whether they know Spanish or not. I had to miss several train connexions causing a delay, but I can say so far that no loss of time has been caused in the work. I suppose that you are looking for good results and not for length of time consumed.*

As there were difficulties in sending funds to Goethe, Lionel de Rothschild approached the Chilean Minister in the Chilean Legation for help. A surprising and welcome outcome was to find that the Chilean Consul in Neuquen, Argentina was a cousin of the Chilean Minister in London who had sent a cable requesting that the Consul do anything he possibly could to assist Mr. Goethe.

Goethe was in the Andes during the period from September 1923 until early 1924 and sent roots and plants back to the Royal Botanical Garden Edinburgh (RBGE) to divide and send out to the subscribers. Henry McLaren wrote to the Regius Keeper, William Wright Smith, of RBGE on 12 June 1924 enclosing a copy of an undated letter he had received that day from Goethe in which Goethe said that he has arrived back in Valparaiso and has collected nearly 300 species which he had sent to the RBGE.

Map of South America showing Beriloche and Neuquen

A further letter from McLaren to Smith on 2 July 1924 said that he had now received notes from Goethe on some 273 numbers of seeds and plants which he is having *transcribed and typed for distribution to Members of the Syndicate.* McLaren goes on to say *The writing is very obscure and illiterate and doubtless there are errors in the transcript, especially as regards the place names, the spelling of which is very hard to decide.* He also says that RBGE should have custody of the originals of the Goethe notes as *it is possible your Staff may be able to correct some of the errors or solve some doubtful points on receipt of the seeds.* An extract of the first page is reproduced below showing the sort of information provided by Goethe.

1. CALLITIA sp. in foothills of Zapala, also almost everywhere in Neuquen. Sandy or often rocky soil; 20%. 3-6ft. flowers not showy.

2. COMPOSITE tall, tomentose shrub 3-4 ft. Fl. yellow like 11. 12.1. stevia; 40%.

3. U. like golden yellow Marguerite or Anthemis. Zapala region. 70% . Miranda Laguna. 11.12.1.

4. 1-3 ft. shrub, evergreen, covered with Heliotrope like flowers with colours varying from white to light salmon, nearly magenta; infloresecence like Verbena. Odour of Syringa, so much so that whole air when passing through the section seems perfumed. Nearly always in loose sandy soil but best colours occur in rocky ground. Only in Section 1 but often in large quantities 900 to 1000 ft. 20%. Leaves short, needle like aculeate.

Information provided by Goethe as transcribed by H. D. McLaren

In a subsequent letter, McLaren sent on to Smith the rest of the transcript of the seeds and plants together with a report by Goethe on the trials and tribulations that he encountered in his expedition. One paragraph summarises the experience:

But taken as a whole it is a rough life full of privations; no one should ever venture into such wilderness who has habits of luxury and is used to much companionship. I do not know of anyone of my acquaintance who would be willing to undergo the task. One has to sacrifice any pleasures of life and often live under most trying conditions as to severity of climate and comfort.

Extract from Goethe's report as transcribed by H. D. McLaren

The papers in the RBGE file show that the seed returned by Goethe was sent to members of the syndicate and also to two other gardens to see which material would germinate in different parts of the UK. A tabulation headed *Andes Expedition Germination of Seeds* has some seven columns headed *Nymans Edinburgh Bodnant Abbotswood Highdown Hidcote Burford.* These refer to the gardens of Lt. Col. Leonard Messel (Nymans – Garden Society member), the RBG Edinburgh (Syndicate member), Hon. Henry D. McLaren (Syndicate member – Bodnant), Mark Fenwick (Abbotswood – Garden Society member), Major F. C. Stern (Syndicate member – Highdown), Major Lawrence Johnston (Syndicate member – Hidcote), and Sir William Lawrence (Syndicate member – Burford, near Dorking). This shows the way in which the keen garden owners of that time – including Lawrence Johnston – would work together with the Royal Botanic Garden at Edinburgh to extend the horticultural understanding and value of the material being sent back by the plant hunting expeditions that were taking place around the world in these years.

Syndicate for H. F. Comber, Andes, 1926 – 27

Some doubt was cast on the quality of what Goethe sent back as a letter two years later dated 25 June 1925 from Lionel de Rothschild to the Chilean Minister says *Two years ago you very kindly gave Mr. Goethe a letter of introduction to a cousin of yours who was at Neuquen. Mr. Goethe sent us back some seeds, but unfortunately he did not understand too well the art of collecting.* His letter went on to say that

> *the same syndicate who sent Mr. Goethe out is now sending a Mr. Harold Comber to try and collect seed in the Andes. Mr. Comber has considerably more experience than Mr. Goethe, having been trained at the Royal Botanic Garden, Edinburgh, and having full cognisance of all Chilean plants growing in our country. Mr. Comber is sailing on the 10th July and proposes to make his way up to Neuquen.*

He then asks that the same facilities that he gave to Mr. Goethe be given to Mr. Comber.

The list of subscribers for the Comber Andes expedition of 1925 – 1926 consists of some sixteen names with Reginald Cory of Duffryn, Lionel de Rothschild of Exbury, J. C. Williams of Caerhays Castle and H. D. McLaren of Bodnant each having two shares of £50 each. Other subscribers included the Hon. Robert James of St. Nicholas, Richmond and F. C. Stern of Highdown, Sussex who were friends of Lawrence Johnston. Although Lawrence is not listed as a subscriber, a letter from Reginald Cory to F. G. Preston, Curator of the Cambridge University Botanic Garden on 24 April 1926 sending on a package of Andes seeds says that Lawrence had accepted Cory's invitation to have a share of these and that some which Comber thinks may not be hardy in England may be successful in his garden in Menton.

I yesterday received the enclosed package of Andes seeds from the Edinburgh Botanic Garden, and I am sending them on to you. I wrote to Mr. Johnston asking whether he would like to have a share of these, and he wrote saying he would be glad of them, and some of them which Comber suggests are not hardy in this country he thinks may be successful in his garden in Italy. I shall be much obliged therefore if you will divide these between yourself and him. His address is:-

L. Johnston, Esq
Hidcote Manor
Campden
Gloucestershire

Reginald Cory letter to F. G. Preston, 24 April 1926

This letter goes on to ask that the seeds be addressed to Mr. Adams who is Johnston's gardener.

Further similar letters requesting a division of the seeds with a half share being sent to Lawrence were written in May and June 1926 and also a year later in July 1927. The records of the Cambridge University Botanic Garden sometimes list the species received from Comber.

Particulars of Seeds collected in Comber from (113)

Discaria No. 420
Chilotrichium rosmarini folium No. 348
Ephedra No. 447 a.
Ephedra sp. No. 447
Habranthus No. 353
Verbena No. 79
Berberis buxifolia No. 417.
Calceolaria No. 82.
Verbena No. 112
Chelotrechium rosmarini folium (No. 348).

Seeds from Comber as received from RBG Edinburgh, 1926

Lawrence Johnston, 1925 to 1928

Lawrence's notebook covering the years from 1925 to 1928 shows that at this time he was visiting gardens both in Britain and abroad to identify and collect species of plants such as *Arbutus canariensis* and *Calceolaria violacea.* His notes show that plants were being sent to both Hidcote and Menton. It confirms his meticulous nature with the recording of itemised details of the costs of travelling on 16 September 1926 from Hidcote to Serre de la Madone – including items such as £9. 9s. 3d for luggage to Menton, 5s 6d for porters on the boat, and then entries in French francs including 100 francs to Maurice for "*passing us through customs*" and, on more than one occasion, 6 francs for grapes. Also recorded is the cost of his chauffeur Ernest Daniel's journey to Menton as being £3 in England and 659 francs in France. Subsequent entries record Christmas boxes to the servants at Menton – of £2 for Ernest Daniels, his chauffeur, and Ellen Graham, who was Gertrude's companion and maid, and 400 francs for the Italian servants.

As already noted Lawrence's mother, Gertrude, died on 8 December 1926 at Villa Mer et Monts adjacent to Serre de la Madone. In 1927, he continued his travels to collect plants for Hidcote and Menton with visits to the Royal Botanic Garden in Edinburgh for *Melaleuca thymoides,* in May to Bodnant, and then to the Chelsea Flower Show for *Lewisia finchae* before calling in at a garden closer to home – Mark Fenwick's garden at Abbotswood for *Paeonia* 'Sunbeam' & 'Fire King'.

South Africa expedition, 1927 – 1928

Then in September 1927, Lawrence Johnston organised a four month plant collecting expedition to South Africa with Collingwood "Cherry" Ingram and Reggie Cory with the newly graduated George Taylor (later to be the Director of the Royal Botanic Gardens at Kew) being paid for by Cory to come as the 'collector'. George Taylor had been a student at Edinburgh University in the Department of Botany which was housed at the Royal Botanic Garden Edinburgh where William Wright Smith was both a professor of botany and the Regius Keeper. Taylor graduated with a first class honours degree in botany in 1926 and then started work on his Ph. D. – it was during his second year that he was invited to be the collector for the Johnston, Cory and Ingram expedition to South Africa.

Collingwood Ingram wrote a journal of the expedition to South Africa with entries for most but not every day, whilst George Taylor wrote a daily diary. Between them, a good impression can be gained of where the group went and what they did. Lawrence together with Collingwood Ingram and Reginald Cory sailed on the *R.M.S. Carnarvon Castle* to South Africa by way of Madeira arriving at Cape Town on 19 September 1927. Collingwood Ingram's journal records that:

> *We had a smooth and uneventful voyage to Madeira which we reached on the early morning of the 6th. The scorching sun of the early hours was tempered later by a mass of grey clouds which gathered round and shrouded the mountain tops. A hurried visit (the boat sailed again at 11 AM) only allowed us to visit three gardens & make a hurried tour of the fruit & fish markets near the quay. The former with its peaches, grenadillos, water-melons, grapes, chillies & other exotic fruits made a fine blaze of colour in the hot sunlight.*

R. M. S. Carnarvon Castle

A good flavour for life on board was given by Ingram's notes:

> *The weather in the tropics has been perfect. A full moon rising as a dull red ball out of the darkening purple of the sea changed to*

radiant silver as she climbed the heavens & throws a quivering path across the waves. A soft wind roars in one's ears. The ship's bows rise & fall in gentle salaam to the moon while the mastheads sweep back and forth through the stars & the small glaucus-white clouds riding silently across the purple splendour of the night.

George Taylor travelled separately and arrived in Cape Town a week after Lawrence, Cory and Ingram. Although not mentioned in Ingram's journal it is evident from a letter from Humphrey Gilbert-Carter, the Director of the Cambridge University Botanic Garden, to Reginald Cory on 2 December 1927 that F. G. Preston had come back from an R.H.S. meeting with *the doleful news that you had hurt your leg on the boat, and had been in hospital ever since your arrival in South Africa. What abominably rough luck.* There is no reply from Reginald Cory until a letter written from Duffryn on 16 March 1928 when he said that he had just returned from his South African tour and says *Many thanks, I am of course quite all right now, and my knee trouble was caused by a blow which kept me rather on the hobble for about six weeks.*

All four members of the expedition appeared to do their own thing to a certain extent as Ingram went off to see the Penguins in Dassen Bay on his own, Reginald Cory was not well at the start and sometimes travelled on his own and George Taylor spent a lot of time at Kirstenbosch working on the collected samples. Lawrence provided the transport – a car and a van – which was waiting for them on arrival in Cape Town and had his chauffeur (Ernest Daniels) and another servant (Fredo Rebuffo) with the group.

During the first month they made a number of trips to visit the wild flower shows in Cape Town as well as inland and to the north – at Malmesbury, at Darling, at Ceres and along the coast to Hermanus. Ingram noted:

We have made excursions in a number of different directions & on each occasion have encountered new & strange plants – many of them of exquisite beauty. Only a fraction of these appear to be in cultivation in the northern hemisphere – & this fraction by no means includes all the best things. The explanation probably lies in the great difference in climatic conditions between England & Cape Colony, which has hitherto deterred English horticulturists from sending collectors to this country. Johnston, who gardens in the Riviera as well as in Gloucestershire, should reap a rich harvest from this expedition as many South African plants are known to flourish on the shores of the Mediterranean.

They also visited Ceres to camp in the Hex River valley on 2 October and went on to the Bokkeveld. The weather turned wet whilst camping as Ingram noted:

A steady tattoo of rain on the roof [of] *our tent awakened me in the early hours of the morning & the drumming increased as it increased into a down pour before dawn. This is the first rain we have experienced in South Africa. By midday the clouds had rolled up & the sun was shining again with its usual brilliance.*

George Taylor's diary for the same day observed: *A wet night & morning. Breakfasted in the Grand Hotel* [in Ceres] *& went to Ceres Flower Show. Then botanised in the mountains behind the camp. Found good succulents, gladioli etc.*

Hex River Camp, Lawrence Johnston seated in the centre with George Taylor on the right. Ernest Daniels standing on the left with Fredo Rebuffo.

By mid-October Ingram recorded in regard to a trip to Elgin in the Western Cape:

> *We have set out again on another round of plant hunting excursions – Johnston, Taylor & myself in one car & the chauffeur & Italian cook-valet in the van with all our paraphernalia. It has been cold all day with occasional flying showers so we were glad to sleep under a roof instead of canvas for our first night. Our mode of procedure is to inveigle some local plant collector (in most "planty" districts there is at least one man who regards himself as an authority & is willing to exploit his knowledge) to come out with us & show us the best places. This is what we did here – a Forest Ranger offering his services as soon as we arrived.*

From Elgin they motored back to the coast and as George Taylor noted they *then proceeded to Hermanus where we put up at the Marine Hotel.* A photograph showed the view along the coast from Hermanus.

The view to the east along the coast from Hermanus

They made an expedition to Table Mountain on 26 October. Another expedition took them over the Garcia Pass and across the Little Karoo to Ladysmith and on to Seven-Weeks Poort by a long winding canyon through the Swartberg Mountains.

Crossing a river in the Swartberg Mountains region

They then started to travel east along the coast and went from Riverdale to George, just east of Mossel Bay, and then on to Knysna. As George Taylor noted on 9 November

> *Left the Wilderness & motored to Knysna. On the way we had glorious views of Kniphofia and saw the malachite honey bird drawing honey from the flowers. At many places we were passing through the Knysna Forest. At Knysna we had lunch & booked rooms at the Royal Hotel. In the afternoon we motored out to Deep Walls*

Forestry Station to select a camp site. This took us through the primeval forest where we saw spoor of elephants. Tree ferns were also seen in abundance.

Knysna Forest, from the left Cory, Taylor and Johnston

They then took the train from George overnight to Port Elizabeth, further east on the coast. From there they continued along the coast to Grahamstown and so to King Williamstown. They then travelled on to Mhtata and thus to Port St. Johns and then via the Magwa falls to Bizana and hence to Durban. As Taylor noted:

> *On to Lusikisiki where we had lunch after visiting the Magwa Falls – 450 ft drop of water highest in S. Africa glorious gorge. Saw tree ferns growing in open veld – ground hornbills, huge birds, loomsday*[sic]. *From Lusikisi on through Flagstaff & on to Bizana where we remained for the night.*
>
> *The country today was much greener & not so much stocked with cattle. The views of rolling hills & distant background of mountains was magnificent.*

Map of South Africa showing places visited in 1927/28

They then went to Eshowe and on to Pietermaritzburg and from here travelled inland to Mont aux Sources & the Drakensberg Range. The journey was clearly difficult as Ingram's journal notes that after an unplanned overnight stay at Winterton because of thunderstorms and heavy rain:

It had cleared by this morning & the roads appeared to be passable, so we continued our broken journey at an early hour. We made good enough progress to Bergville but beyond that our way was fraught with difficulties. Our tyre chains came off; we repeatedly stuck on the greasy slopes; we twice walzed round on to the veld &, on one occasion, our engine was quenched in the midst of a flooded spruit. Poor Fredo (Johnny's Italian servant) was a little perturbed at the car's misbehaviour "Doucement, doucement, il faut marché avec grand prudence" he kept advising. We finally covered the 40 odd miles to our destination (The National Park Hostel) at Mont aux Sources.

Mont aux Sources, Drakensberg Mountains

Ingram was clearly impressed by the Drakensberg as he wrote:

When one stands near these immense escarpments & becomes fully conscious of their towering height they give one the appearance of walling in both sky & earth; with a canopy of cloud resting upon their edge they might well be a banner forming the world's end. I have never seen mountains quite like them – they have a character all of their own, with their smiling slopes & terrifying cliffs.

They visited the Tugela Gorge & Falls and also rode up to Sentinel Cave at about 10,000 feet, spending the night there in the cave, and visited the summit plateau, Ingram noted that:

As these caves are about 10,000 ft above sea-level they were none too warm & despite an excellent fire kindled by our Zulu guides & a large number of rugs, we found the night unpleasantly damp & cold. The floor too was hard & unyielding. As I lay awake listening to the hollow drip, drip, drip of falling water, trying to conjure up thoughts to while away the endless hours, I became firmly convinced of the undesirability of a troglodyte's existence. At last the triangular opening began to pale with the dawning of day & we found that the rain of over night had ceased & that there was promise of fair weather to come.

Climbing through a narrow pass on to the upper plateau I found myself in brilliant sunshine above the clouds. These lay like a shining sea of billowing folds below me, pierced here & there by the jagged escarpments of the blue-black mountains. On the westward side the clouds had broken up in to a series of islands & these had floated away still on the same level plain, across the grassy flanks of a lower & distant range until they finally became lost in the halcyon blue beyond.

View from Mont aux Sources

From here they went on 6 December to Ladysmith and then via Johannesberg to Pretoria. Whilst in Pretoria, on Sunday 11 December they visited General Smuts, a keen botanist, on his farm south of Pretoria. George Taylor's diary entry records:

General Smuts Day.

Motored out to Irene in the Government caravan & first of all picked up Dr. & Mrs. Pole Evans. Then we went to Gen. Smuts farm where he showed us round the farm & his house. He explained various mementos – the flag of the Transkei Republic, ...

Motored out on low veldt & had a sumptuous lunch including various drinks & ice cream!

Coming home to Irene where we had tea with the Smuts family we saw a huge herd of Blesbuck [Blesbok], *also saw Springbok & Roibuck* [Reebok].

• *Exceedingly interesting & glorious day.*

From the left, I.B. Pole Evans, George Taylor, Liebenberg, General Smuts, C. A. Smith, Daltonate, Lawrence Johnston, Mrs. I. B. Pole Evans and Reginald Cory.

Ingram was equally impressed and recorded that:

Today Dr Pole Evans motored us out to Gen Smuts farm at Irene, some ten miles or so from Pretoria. Notwithstanding his military & political activities Smuts has found time to acquire considerable botanical knowledge & he has made a special study of S. African grasses. A family man, obviously honest to his own convictions & with a highly-trained, well-balanced mind, I found in him a very attractive personality who conversed freely & openly upon any subject that was broached.

From Pretoria they travelled by train north into what was then known as Rhodesia (now Zimbabwe) to visit the Victoria Falls returning through Bulawayo to Pretoria for Christmas Day 1927.

Victoria Falls, December 1927

Ingram in his journal noted that:

The falls are undoubtedly one of the wonders of the world: to my mind infinitely more beautiful than Niagara being set in wilder &

grander surroundings. The native name of "Mosi-ou-tunya" is said to mean "The smoke which sounds". If so, one cannot help regretting that such an apt & poetic appellation has not been maintained. From afar one sees a great steam-like cloud rising in a white column from the forest & one's ears are filled with the dull booming roar of falling waters.

Whilst at the Victoria Falls, George Taylor noted that *Major Johnston & I had a 12 mile walk – through bush to the Masuyi River Falls. Wet & dry we got alternately.* On the journey back to Pretoria, Ingram notes that they went to visit *World's View*, the burial place of Cecil Rhodes, in a notable area:

The interest lies in the remarkable & fantastic rock formation, many of the kopjes being formed by rounded shoulders of bare granite, while over the whole landscape are piled titanic boulders in all manner of odd & un-natural positions – looking like Brobdingnagian pebbles that had been playfully scattered by some stupendous & super-human agency.

World's View, Matopos, burial place of Cecil Rhodes

Ingram went on to add that:

> *I was told that a pinkish-red Gladiolus was to be found in the vlei, or grassy valley, through which the road passed, but owing to the dry season there were no flowers to be seen. However, by diligently searching I was at last able to detect a few plants that were growing amongst the very similar grass-like vegetation. I consider myself fortunate in being able to find it under such conditions. "I believe Cherry must smell them out" said Johnnie.*

Following their return to Pretoria for Christmas, Collingwood Ingram left the rest of the party on 28 December and returned to Cape Town and then on 6 January 1928 sailed back to England on the *Walmer Castle* via the West Coast route. His journal on 30 December noted: *Arrived at Cape Town this morning having "wept a fond farewell" of the others on the 28th. It is their intention to do a little more collecting in Transvaal & then return by the East Coast route – a programme which will take at least a couple of months whereas I am due in England on Jan 23rd.*

The others made visits to places such as Barberton in the area around Pretoria and then on 10 January went to Pietersberg, North Transvaal returning to Pretoria on 16 January before Lawrence and Reginald Cory two days later left Johannesberg for Lourenco Marques. They then returned by the East Coast route, probably on the *Llandovery Castle,* which sailed from Cape Town calling at Algoa Bay (Port Elizabeth), and Natal before Lorenco Marques and then at Beira, Dar-es-Salaam, Tanga, Mombasa, Port Sudan, Port Said and Marseilles before arriving in London on 22 February 1928. As neither Lawrence or Reginald Cory are listed amongst the incoming passengers arriving in London, it is likely that they disembarked at Marseilles in order to visit Serre de la Madone. They probably arrived back in England at the same time as Fredo Rebuffo whose Registration Card records *Arrived at Hidcote Manor from South Africa via Dover on the 4th April 1928.* Several of the genera recorded in Lawrence's notebook during the time that he was in South Africa if not the species -- such as *Aloe, Cotyledon, Cyrtanthus, Felicia, Hypoxis, Ipomea, Jasminum, Kniphofia* and *Lobelia* are still flourishing at Hidcote.

George Taylor went back to Cape Town and Kirstenbosch until he left Cape Town on the *Llanstephan Castle* on 3 March 1928 also sailing back up the East Coast and through the Suez Canal.

Collingwood Ingram's journal and George Taylor's diary together provide an interesting insight into this plant hunting expedition. George Taylor's additional technical knowledge was also recognised in a note about Dr. Muir, a keen botanist, of Riversdale, inland and to the east of Cape Town:

> *Yesterday we motored over Garcia Pass & across the Little Karoo (which has been suffering from a terrible drought) to Ladismith & from there on to the famous Seven-Weeks Poort. Dr Muir accompanied us. I have never known a man so fond of words! He literally revelled in the longest & most sparsely used botanical terms & rolled them off his tongue with all the gusto of a connoisseur tasting a fine vintage port. Taylor could stand up to this many-syllabled cascade but the colloquy was rather too technical for Johnny & myself. However, Muir knows his plants well & was of great assistance to us.*

George Taylor's diary shows that he was always careful to refer to Lawrence Johnston, Reginald Cory and Collingwood Ingram in his diary as Major Johnston, Mr. Cory and Mr. Ingram respectively or by their surnames. And as the collector, he stayed on at Kirstenbosch for some six weeks working on what had been collected. The other insight is that the expedition included calls on landowners such as Sir Lionel & Lady Phillip at Vergelegen near Somerset West and involved stays at hotels such as the Marine at Hermanus as well as camping.

Mount Kilimanjaro, February 1929

Early the following year, in February 1929, Lawrence Johnston's diary showed that he travelled to Kenya to climb Mount Kilimanjaro. It seems likely that Lawrence had taken the opportunity whilst sailing back from South Africa in January 1928 to make inquiries at Mombasa on how best to reach Mount Kilimanjaro. Certainly the first two entries in the 1929 diary in spaces prior to that for 1 January are of names of individuals -- *R. S. Campbell, Vasco da Gama St., Mombasa* and *W. W. Samuel, Mombasa.*

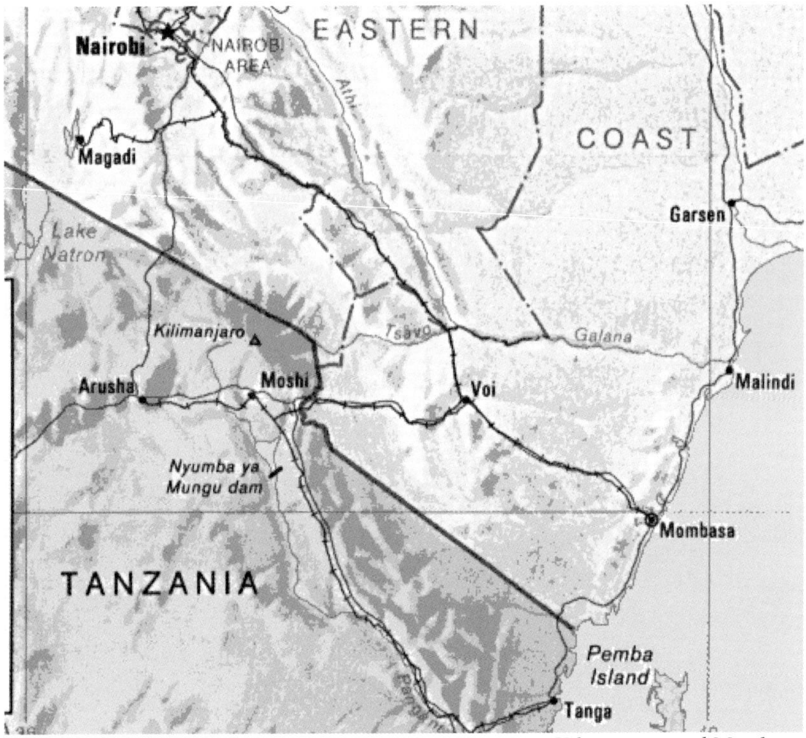

Map showing Mount Kilimanjaro and Mombasa

Lawrence's diary shows that on 11 February he motored from Nairobi to Marangu just to the west of Moshi on the slopes of the mountain:

> *Motored from Nairobi*
> *to Marangu. Saw all sorts*
> *of game and a great gathering*
> *of storks & locusts*

The next day he visited Moshi and then came back to Marangu. From here, he climbed to the Bismarck Hut (which still stands today alongside the present Mandara Huts) at about 2,600m on the side of the mountain. His diary records that he *Went to top of the forest* and *gathered seed of Impatiens violet* and *Senecio.* In some of the very rare descriptive passages in his diary, the entries for the following week describe climbing on the mountain.

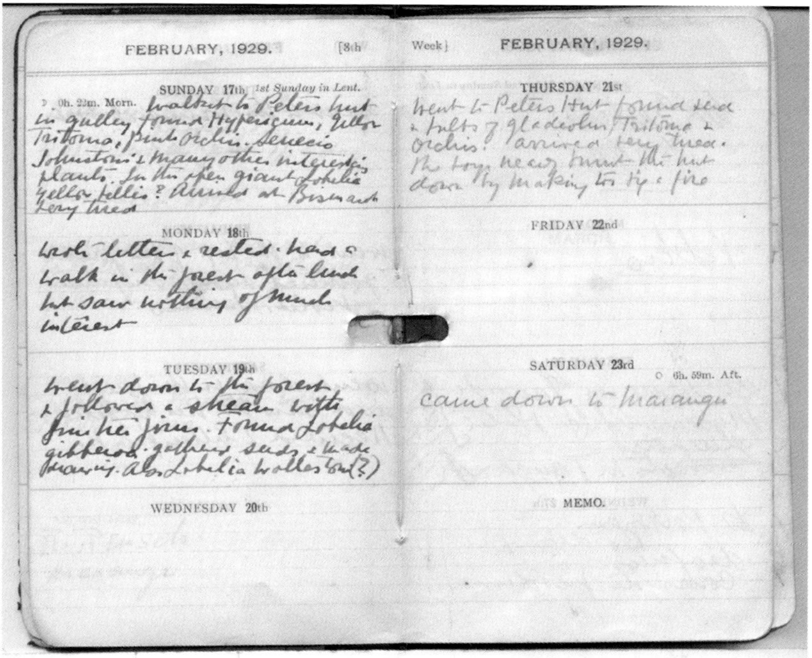
FEBRUARY, 1929. [8th Week] FEBRUARY, 1929.

SUNDAY 17th 1st Sunday in Lent. 0h. 22m. Morn. walked to Peters hut in gulley found Hypericum, Yellow Tritoma, pink orchis. Senecio Johnstonii & many other interesting plants. In the open giant Lobelia yellow fellis? Arrived at Bismarck very tired

MONDAY 18th wrote letters & rested. had a walk in the forest after lunch but saw nothing of much interest

TUESDAY 19th went down to the forest & followed a stream with fine tree ferns. found Lobelia gibberoa. gathered seeds & made drawing. Also Lobelia Wollastonii(?)

WEDNESDAY 20th

THURSDAY 21st Went to Peters Hut found seeds & bulbs of gladiolus, Tritoma & orchis. arrived very tired. the boys nearly burnt the hut down by making too big a fire

FRIDAY 22nd

SATURDAY 23rd 6h. 59m. Aft. came down to Marangu

MEMO.

Lawrence Johnston diary, 17 – 23 February 1929

On 17 February he walked to Peter's Hut (now known as the Horomb Hut) located at about 3,700m and noted that *in the gully found Hypericum, Yellow tritoma, pink orchiis, Senecio johnstonii.* Was this the *Hypericum Hidcote* that now flourishes at Hidcote? Four days later Lawrence records *Went to Peter's Hut – found seed and bulbs of gladiolus, tritoma & orchis. Arrived very tired. The boys nearly burnt the hut down by making too big a fire.* Two days later he descended to Marangu.

This was the trip Lawrence described in the only article he is known to have written entitled *Some Flowering Plants of Mount Kilimanjaro* published in October 1929 in *New Flora and Silva.* His writing is evocative and gives a good sense of the experience:

After all, I realized that in all that waste of dry grass and scrub and even of tropical forest, where lush green leaves prevent the growth of many flowering plants, Kilimanjaro has its little flower gardens, hidden in rocky dells, sheltered from wind and watered by streams from the melting snows. The King of these little valleys is Senecio Johnstonii - named after Sir Harry Johnston - as picturesque a plant as so essentially an ugly one can be. It looks like a huge Artichoke of pale green leaves on a woody stem, sometimes as much as 15 feet high, and often branched into several heads. The old leaves, dry and brown, hang down and cling to the stem, making a thick mantle from which the brilliant green tufts of leaves arise. This colour tells tremendously against the neutral background of the rocks and grass. It often grows on a high boulder or a mound with great effect, as if conscious of its kingly dignity. The plant was not in flower, but I secured a good supply of seed, and I hope that, at least in some places, it will prove hardy in England. It goes very nearly up to the limit of vegetation on Kilimanjaro, where, even in the hot season, it is subject to almost nightly frosts. I only saw it growing where its roots were within reach of moisture and where the plant was comparatively sheltered from wind.

Every step we took over the black boulders of this little gully brought some new treasure into sight. The banks were covered in Heather and long grasses, and many Ferns were about the shaded pools. A very fine dwarf Hypericum, with slender foliage and large, deep, orange-yellow flowers with dashes of red, grew amongst the Heather, and may be hardy in our gardens, as this, too, grows at a great altitude and always near where streams had dug their beds deep into the rock and there is moisture and shelter and hot sun.

There were many varieties of Helichrysum, one with brilliant pink flowers, and several others with grey foliage and yellow flowers. As I came around a bend of the stream there was a tall clear-yellow Tritoma with glaucus leaves showing up brilliantly against the black lava cliff. I think this is only a colour variety of the one growing lower down, for I found it afterwards in all shades from yellow to red, but this form looked especially lovely and delicate in its dark surroundings.

In one place, where there was moisture and shade, were large clumps of a tall bright pink Orchis with many spikes in full bloom, but luckily with some ripe seed. I hope that plants will be raised from seed of this which I sent home. It is as big as Orchis foliosa, but a brilliant pink and, to my mind, a much more lovely thing.

The entries in Lawrence's diary do not show what he did after his expedition on Mount Kilimanjaro although the names and addresses suggest that he had returned to France and by the end of April he was back at Serre de la Madone as he visited La Bollene inland from Menton.

George Forrest expedition to Yunnan 1930/1931

By 1930/1931 George Forrest had made some six plant hunting expeditions. The first and second were funded by A. K. Bulley of Bees Ltd and the third by J. C. Williams of Caerhays Castle, Cornwall in the years before World War I. Then after the war, his expeditions were funded by syndicates of garden owners in which Reginald Cory of Duffryn and J. C. Williams were prominent. Lawrence Johnston will have heard of George Forrest and his expeditions from Reginald Cory and from other members of the Garden Society.

On his return from Kenya, Lawrence wrote in April 1929 to William Wright Smith, the Regius Keeper at the Royal Botanical Gardens Edinburgh, expressing his interest in sponsoring a plant hunting expedition by George Forrest. Although Johnston's letter is not in the RBGE file, Smith wrote on 11 April 1929 to George Forest saying:

Herewith a letter from Lawrence Johnston which will interest you. He is evidently very keen and quite willing to make the journey up here if in any way he can advance arrangements for an expedition.

George Forrest replied on 14 April saying:

It was good of you to send Mr. Johnstone's [sic] *letter to me & I herewith return it. Many thanks! He certainly seems keen & I suppose May 15th would be convenient for a meeting, but we shall settle that when I see you on my return on Saturday the 27th.*

It seems likely from entries in Lawrence's diary for 1929 and the correspondence with Forrest in the RBGE files, that there was a meeting in Edinburgh in May 1929 between Forrest and Lawrence. The ensuing correspondence between Smith and Forrest in the RBGE files which are incomplete shows Smith wrote on 13 August saying *I have at the same time another letter from Lawrence Johnston which I enclose. Stern's letter referred to therein is only a short one. Herewith is the only important part in it:-*

> *"It entirely depends how much the expedition is going to cost whether the money can be raised or not; and secondly it is essential that J. C. Williams and Rothschild should be consulted before anything is made public or subscriptions are asked for. I am going to America in September for 6 weeks."*

Smith continues by saying *I quite agree with the general drift of the two letters.* He then considers how best to seek the views of Williams before concluding by saying *What do you think yourself? In any case I think you should let Major Johnston have a line from you.* George Forrest replied to Smith in a letter on 31 August 1929 saying:

> *Re. the proposed expedition & Mr. Johnston. After all that has transpired and having given the matter much consideration from every angle I have decided to have nothing more to do with the affair, at least as far as the present proposal is concerned & I am writing Mr. Johnston informing him of my decision.*
>
> *I have no intention criticising either Johnston's or Stern's action, or what the latter said as I know it would be useless. I have not forgotten, nor shall I ever, what happened last year with J. B. Stevenson, how I did my utmost to discuss the cause of the trouble, & how I failed. If such people prefer to act in an ungentlemanly manner it doesn't affect me much, only in that I shall refuse to work for them. Certainly I do not gain, but still they lose, or at any rate that is as I see it.*
>
> *I never had the slightest trouble in my dealings with the promoters of any of my past expeditions & I certainly do not intend entering any agreement now where there is likely to be friction. It's not worth it.*

Besides as you know I have never favoured an expedition where there is likely to be 40 – 60, or maybe more small subscribers, such as Stevenson suggested and now Stern proposes. I think you will agree with me in that. I have always been accustomed working for a few & I have found it most satisfactory, & when I desire I think I shall be able to arrange it so again. Despite Stern's talk, I consider that I am the best judge of my condition & I feel I shall remain for quite a few years yet fit enough to take up the work if necessary.

I regret if my decision should disappoint Johnston. Still I have no doubt he may get someone to take my place. I hear Ward is on his way home& I'm sure he'd be willing to do the stint at a price!

There is then a gap in the correspondence in the RBG Edinburgh files. However, Lawrence's diary has an entry on 30 September *go to Scotland* and then on the following day *10 o/c 17 Inverleith Place* – the Edinburgh address of George Forrest. A further entry for 12 October shows *Meet J. C. Williams at the North British Hotel at 8.30 a.m.* In view of the importance attached to the views of J. C. Williams in the Smith letter of 13 August about Major Stern's views it is likely that this meeting at the North British Hotel – which is in Edinburgh – was to discuss the proposed expedition. Two days later on 14 October when Lawrence returned to Hidcote he wrote to Henry McLaren at Bodnant saying:

I have just returned from Edinburgh & have quite decided that it would not be of much use for me to try to go to Yunnan alone & Forrest still thinks he is going there next year.

Some three days earlier, on 11 October, George Forrest had written to Henry McLaren saying:

Re. the expedition planned for next year. Professor Smith & I have decided that shares be £500, and eight or ten at that will be necessary. Putting it at that figure does not, of course, mean that five – more or less – persons may not club to take a share.

However, you will probably meet Professor Smith at the meeting of the Garden Society on the 19th when he shall explain everything to you, and others, who may wish to participate.

A further couple of weeks later, on 29 October 1929, Smith wrote to Lionel de Rothschild saying:

> I have had very much in mind for some time the possibility of a final Forrest expedition into West Yunnan and Tibet. It is after all the richest area in the world for horticultural Europe. It seems to me, too, to be the last opportunity of utilising the extensive knowledge of the area possessed by Forrest. I do not think that in our time there will be any successor in these regions. The expedition would probably be planned to give two seed seasons and one flowering season. The chief object would be to secure seed of all the species of horticultural merit while botanical material would be entirely subordinate. All Forrest's trained men are still at his disposal and the results would be in correspondence. Rhododendron would not be neglected but it would not be stressed. The various ranges of Yunnan are sure to yield a rich harvest.
>
> *W.W. Smith letter to Lionel de Rothschild, 29 October 1929*

Lionel de Rothschild replied the following day to say *I should be pleased to subscribe for a couple of shares in the new Forrest expedition but do not feel disposed to finance the whole expedition myself.* Lawrence wrote on 10 December to Smith enclosing a letter he had received from Sir Stephenson Kent in which he says *Thank you for your letter of the 4th.* [December 1929] *Certainly I will subscribe £100 to George Forrest's expedition.* Lawrence goes on to ask Smith to write to Mr. H. S. Harrington of Dunloe Castle, Killarney as Lady Moore, the wife of Sir Frederick Moore, the Director of the Royal Botanic Gardens Glasnevin, had suggested that he might be interested in subscribing. Lawrence makes it clear that he had written to Lady Moore to inquire if there are any other likely subscribers besides Lady Londonderry of Mount Stewart and Lord Headfort of Birr Castle. His letter continues *Lady Londonderry will subscribe £50 but I hope to bracket her with someone else for £50 & I hope to get £100 from Headfort. I am becoming very mercenary!*

It is thus evident that Lawrence was very actively engaged in helping to raise the subscriptions for the expedition.

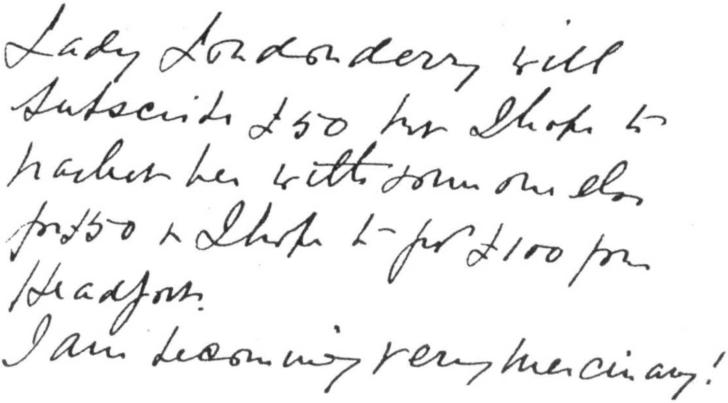
Lady Londonderry will subscribe £50 but I hope to match her with some one else for £50 & I hope to get £100 from Headfort.
I am becoming very mercenary!

Extract from Lawrence Johnston letter to W. W. Smith, 10 December 1929.

The following month on 27 January 1930, Smith wrote to Major F. C. Stern saying

> *As you know we are all trying to fix up another Forrest expedition for two seed-collecting seasons. The aim is a sum of £4000. Major shares to be £500. So far we have promises of £3000, made up of 3 at £500 and the others running from £100 to £125. I think the remainder can be secured.*
>
> *I am enclosing two statements – one giving an outline of the area and one stating the genera likely to be secured. There is no intention of making this expedition confined to Rhods.* [Rhododendrons] *Seed-collecting is the main object. Mr L. Johnston is accompanying Mr Forrest.*
>
> *We have but recently started securing subscriptions and the response so far has been most encouraging. It will be Forrest's last venture. We shall be delighted to have you on the list if you would care to consider it.*

This is the first indication that Lawrence would be accompanying George Forrest. This hardly seems surprising given the enthusiasm shown from the outset by Lawrence and his involvement in helping to raise the funds for the expedition and that he was one of the four who contributed £500, as did Lionel de Rothschild, J. J. Crosfield and the Royal Horticultural Society.

A letter from George Forrest on 1 May 1930 to Smith *encloses a list of paids & nonpaids. I send it in case we may get mixed up. All paids have been acknowledged (except Col. Stephenson Clarke just received).* The list of *Paids* of some 19 names is clearly in order of receipt as it starts with Reginald Cory (£250) and is not in alphabetical or amount order as it continues with H. D. McLaren (£125), J. C. Williams (£200), J. J. Crosfield (£500), Sir Stephenson Kent (£100), Major Lionel de Rothschild (£625) and ends with A. K. Bulley (£125) and Col. Stephenson Clarke (£125). The list of *Yet to pay* is made up of some eight names headed by Major L. Johnston *to pay in Rangoon!* (£500) and including Lord Headfort (£100 or £125).

A later tabulation of the subscriptions shows that the total cost of the *Forrest Expedition 1930 – 1932* was to be *£5000* made up of *10 lots (each £500) + 1 small one for Edinburgh + E. H. M. Cox (£100) + F. D. S. Sandeman (£100).* There were four such £500 lots from Lionel de Rothschild, J. J. Crosfield, the R.H.S. and Lawrence Johnston respectively and then two lots of £500 made up from two sets of 4 subscribers each paying £125 and then two lots of £500 made up of two sets of 5 subscribers each paying £100 and the final lot of £500 made up of 8 subscribers each paying between £25 and £100. There were thus some 32 subscribers. In the end the number of subscribers increased to 39.

The expedition to Yunnan, China, began with Forrest leaving Edinburgh on 7 November 1930. The route followed was to sail to Rangoon, Burma and then to travel inland to Bhamo, upper Burma and there to cross into China and to travel on to Tengyueh in Yunnan Province. It is evident that Lawrence fell ill as a letter from Smith to Forrest on 12 February 1931 includes in its penultimate paragraph that *News of Major Johnston's illness has percolated through as I had a note the other day from Lady Leconfield who called it double pneumonia. I hope to have reassuring news of him.*

Later the same month on 26 February 1931 a letter to Smith from George Forrest, who by then is at Tengyueh, Yunnan, starts by apologising for his long silence and then goes on to express his dissatisfaction with his travelling companion in no uncertain terms. He says in regard to his silence that *really I am not to blame as since my arrival here I have had a most harassing time, even worse than my experience in Rangoon & Bhamo.* He goes on to say that he had last written from Bhamo on 3 January and adds *& that letter would give you some slight idea how badly things had been going. But even then I only gave you a bare outline of what had happened as I had no wish to put too much on paper.* There appears to be no copy of this 3 January letter in the RBGE archives. Forrest in his 26 February letter goes on to say:

> *Had I raked G. B.* [Great Britain] *with a small tooth comb I couldn't have found a worse companion than Johnston and I cannot say how often during the past three months I have cursed myself for being so foolish in consenting to him accompanying me! I have indeed paid for my folly!*
>
> *Well; to carry on from the date of my Bhamo letter. Johnston had left me to take up his residence with the Clerks, leaving me – as he did in Rangoon – to attend to everything in the way of preparation for our journey – as engaging chairs & chair bearers, coolies, & mule transport, etc. I only had an Indian boy as help & as my Chinese was a bit rusty I had a hard row to hoe & was well put to it at times. There was much he could have done to lighten my labours, but he was too busy gadding around with Mrs. Clerk & others all & every day, riding in the morning, tea & tennis in the afternoons & bridge at the Club in the evenings. Knowing me as you do, you may not believe it, but I was more than patient under all of it, but at last I did give way after I hadn't seen him for 3 days. I sent for him & asked him what he meant by all of it & if he thought I was a Cook's courier arranging a tour for him & if he thought I was paid to attend to him. However my effort was fruitless. I might as well have been silent as he never even apologised! For another 2 days he vanished & then the Clerks sent over to inform me he had fallen seriously ill. Apparently he had contracted a severe chill through exhausting himself in playing tennis & then sitting cooling off instead of changing. The Bhamo Divisional surgeon was called in to attend him & after running the tape over him, he informed me that Johnston was in a very bad state internally, chronic congestion of the liver, lungs exceedingly weak & heart & kidneys bad & that it would be extremely dangerous for him to attempt such a journey into Yunnan as we had contemplated.*

> *Johnston then decided he wouldn't go & as I had everything prepared to set off on the 4th Jan., I had another hectic day cancelling all arrangements I had made for him. However, I adhered to the plans made for myself and set out for Tengyueh on the 4th Jan arriving here on the morning of the 10th.*

George Forrest then describes how he found his men awaiting him with a collection of herbaria numbering nearly 1000 species with seed of some 3 – 400 of them. Forrest then went down with a sharp attack of ptomaine poisoning on the day after his arrival. He then continues:

> *Then to add to my discomfort on the 15th I received a wire from Bhamo informing me Johnston had, despite all advice, left on the 14th to come on here. Well! He got in on the 22nd & owing to the strain of the journey immediately became sick once more, an attack much more serious than that he had had at Bhamo.*
>
> *Fortunately there is a very capable Burman here, a hospital assistant lent by the Indian service to attend the Consul and Commissioner here, else I fear poor Johnston would have gone under for keeps.*
>
> *Apparently there is Bright's disease in his family, a fact not divulged to the Bhamo Dr & the trouble developed seriously during this his second illness. Dr Sun Hline gave him a thorough overhaul & managed to fix him up, but informed me that under no circumstances should he be allowed to proceed further, as his kidneys and other organs were in such a critical condition that given such another attack he might die suddenly.*
>
> *Apparently at last Johnston realised the seriousness of his condition & decided to give up & return home! I understand he has therefore arranged to sail from Rangoon about mid April & will go straight to Mentone* [sic], *but as he changes his mind more frequently than his socks I give no guarantee what he shall eventually do. The only point I am quite positive about is that he shall not travel with me!*

Forrest's letter refers to Johnston having Bright's disease – kidney disease today known as nephritis – from which his grandfather, Lawrence Waterbury, had died.

A letter to Forrest from Smith on 11 March, written before he received Forrest's letter of 26 February, says that he was *glad that everything is going on well and that you are very pleased with the collection your men have made during the preceding season.* The letter goes on to say:

> *I had also two short letters from Major Johnston expressing his disappointment but writing, I thought in very good spirits and with a very commendable spirit of resignation. He talks of looking in at Peradeniya on his way back and I think he should certainly take the opportunity of doing so. By the time this reaches you he will have left for Rangoon. I hope you were able to forward him comfortably and that at any rate he was able to see some of the treasures of Yunnan before he had to leave.*

Peradeniya is located close to the city of Kandy on the island of Sri Lanka (then known as Ceylon) and is famous for the Royal Botanical Garden Peradeniya, renowned for its collection of more than 300 varieties of orchids.

The next letter from Forrest to Smith in which he mentions Lawrence is dated 31 March and was again written at Tengyueh. This says that packages of seeds and cases of herbaria have been sent. Forrest then in some ten pages outlines his expedition from 9th to 22nd March and says he intends to leave next week to journey to Lichiang and the north. Forrest then goes on to write about Johnston saying:

> *However, I wish to see Johnston set off on his journey home before I proceed on mine! I regret to say he has been very ill again though sufficiently recovered to allow of his leaving on the 11th or 13th April.*
>
> *He wanted to accompany me to the Ho-tou – the Upper Shuichi – but the Dr. prohibited that & then foolishly he decided on a journey to Yung chang fu against both the Drs and my advice. He went. The trek proved too much for him & on his return journey, two or three days from here he suddenly became seriously ill & sent a runner in with an S.O.S. to the Dr. begging him to go to his assistance which he did & just in time. Later Johnston told me that he thought he was*

really going to die.... It must have been a very close call. Kidneys, the Dr. said.

Forrest continued to say that *a week ago he* [Johnston] *set off for the hot sulphur springs, which lie a few miles south of here, & has been having frequent & prolonged baths. Two days ago the Dr. visited him & reported great improvement. I understand he returns today.* Forrest then adds that *He is a most obstinate person & if he doesn't get away from here next week he may stay here for keeps!*

The next mention of Lawrence is in a letter of 14 May from George Forrest's wife to Smith in which she said that George had written as he was setting off for Lichiang and that *He had seen Mr. Johnstone* [sic] *off on his way home, before he left.*

Smith writing to George Forrest on 3 June says *I have heard nothing more regarding Major Johnston but I suppose he will be arriving in the South of France at no distant date.* Another letter two weeks later on 16 June 1931 says *I dare say Major Johnston is now somewhere in France but I have heard no word of him as yet. He may have stayed longer in Ceylon than he intended. That is likely for otherwise I should have had word from him ere now.* George Forrest continued with the expedition but early in 1932 on 5 January died near Tengyueh.

It is likely that there was a fundamental misunderstanding on the part of Forrest as to what his role was and why Lawrence was there. Lawrence would have expected Forrest to do the field work for which he was being paid whilst Lawrence, having had a major role in raising the funds for the expedition, wanted to go to see how it all went. In other words, they were the employee and employer. Forrest's expectation that Lawrence would be engaged in the day to day work of the expedition, was unrealistic especially given the social circumstances of the 1930s. Lawrence was a member of the Garden Society made up of wealthy landowners who were keen gardeners of whom several had subscribed to the expedition whilst Forrest was the plant hunter who was being paid to carry out the expedition. It will be recalled that it is also evident from the earlier correspondence in the RBGE archives that there had been an incident the preceding year in regard to J. B. Stevenson, another member of the Garden Society, that led to Forrest writing that *If such people prefer to act in an ungentlemanly manner it doesn't affect me much, only in that I shall refuse to work for them.* It is also evident from the files that Forrest could be somewhat peremptory in his correspondence with his sponsors. Lawrence had also had the earlier experience of his plant hunting expedition in South Africa in 1927 – 28 when George Taylor was the collector and there had been no such difficulties.

A photograph has recently been found by RBG Edinburgh showing Lawrence on this expedition.

Lawrence Johnston on the Forrest expedition, 1930-31

Magnolia macrophylla from the U.S. Capitol, Washington, D.C.

On 10 November 1931 B. Y. Morrison, the Senior Horticulturist of the Bureau of Plant Industry in the U.S. Department of Agriculture in Washington D.C. wrote to Lionel de Rothschild saying that he was sending some six ounces of seed collected from the *Magnolia macrophylla* trees in the grounds of the U. S. Capitol Building in Washington D.C.

> So many people in England seem enormously keen to have seed of Magnolia macrophylla that I am taking the liberty of sending you about six ounces of seed collected from the trees here in Washington on the grounds of our National Capitol Building.
>
> *Morrison letter to Lionel de Rothschild, 10 November 1931*

The letter invited Rothschild, if he wished, to share the seed with his friends. It is evident that the letter is in response to Lionel de Rothschild having shared some of the seed from the Forrest expeditions with Morrison.

Rothschild replied on 23 November thanking Morrison for the seeds and saying that he proposed to distribute them *amongst the members of the Garden Society and the Rhododendron Society who may want it, as I think you met all of these gentlemen.* A week later, on 1 December, he wrote again listing the 27 recipients – which included Mr. L. Johnston – as well as adding that *I am sowing some at Exbury.*

It is interesting to note that among the plants listed as being in cultivation at Hidcote in the 1950s in Appendix D to Katie Fretwell's report of 1999/2000 on Hidcote is *Magnolia macrophylla* although this is no longer growing at Hidcote in 2010.

Yashiroda expedition to Formosa (Taiwan) 1932

Lawrence's experiences in Yunnan had clearly not put him off plant hunting as later in 1931 he wrote to Kan Yashiroda of Tonosho-Kyuku, Kagawaken, Japan to inquire into Yashiroda's plant hunting expeditions. When he wrote on 1 January 1932 to Henry D. McLaren to see if he would be interested in taking a share in a proposed syndicate to fund an expedition to the highlands of Formosa (now known as Taiwan), Lawrence makes it clear that he had been introduced to Yashiroda by George Simpson-Hayward of Icomb Place – another gardener who lived some 15 miles south west of Hidcote on the other side of Stow-on-the-Wold. Kan Yashiroda trained at the Royal Botanic Gardens at Kew for a year in 1925—26.

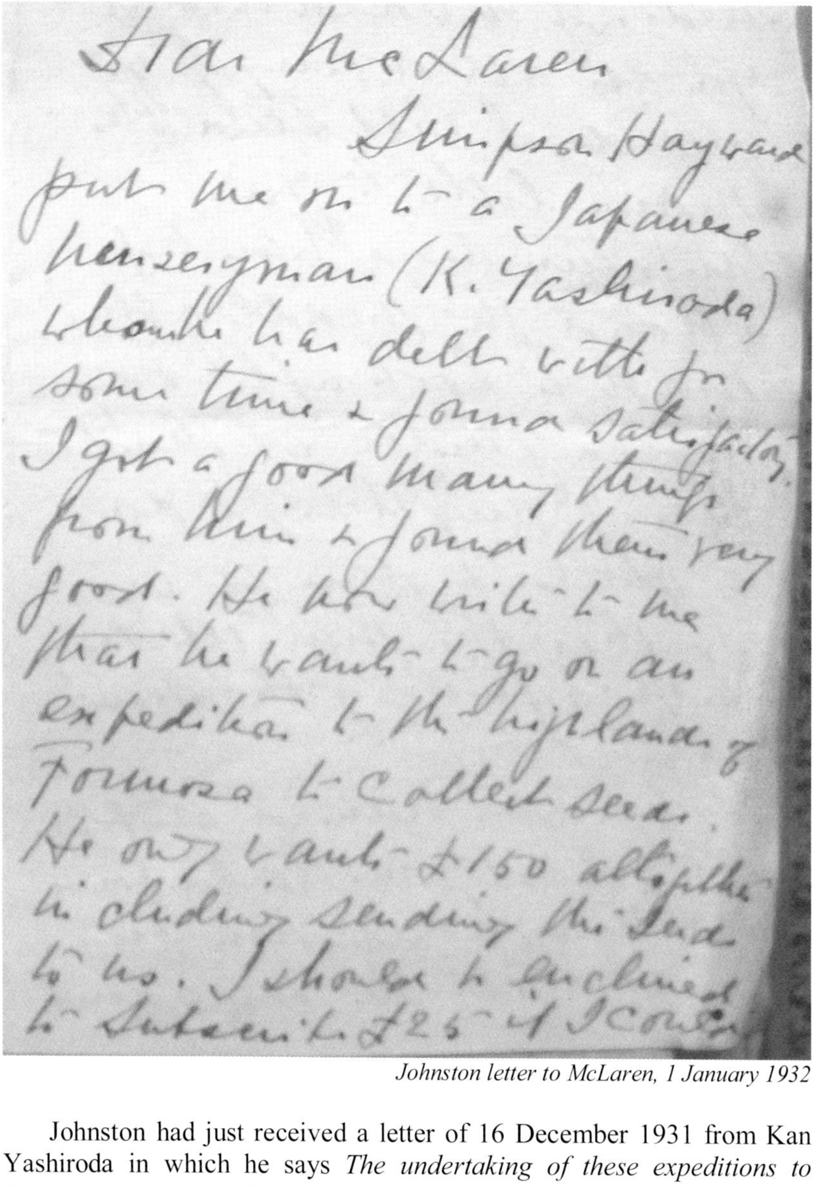

Dear McLaren
Simpson Hayward
put me on to a Japanese
nurseryman (K. Yashiroda)
whom he has dealt with for
some time & found satisfactory.
I got a good many things
from him & found them very
good. He now writes to me
that he wants to go on an
expedition to the highlands of
Formosa to collect seeds.
He only wants £150 altogether
including sending the seeds
to us. I should be inclined
to subscribe £25 if I could

Johnston letter to McLaren, 1 January 1932

Johnston had just received a letter of 16 December 1931 from Kan Yashiroda in which he says *The undertaking of these expeditions to Formosa and the Kurile Is. etc., is not, I think, a task but an experience in life, to the youth whom are ambitious to bring out to the horticultural world as well as the science some new or meritorious plants.* He goes on to outline the plants to be found at various heights in Formosa. He then describes how the plant hunting would be carried out giving a graphic description of some of the dangers involved:

> tribe
> I collect employing two or more men of Takasago/ (ferocious aborigines living on the highlands and high mountains throughout F ormosa whom are skilful in climbing on the lofty tree: these savages killed more than 200 people two years ago and still occasionally: they proud very much that their houses are decorated by many craniums of the foreign who, were killed by
> But such a matter is nothing to me and I can manage.
> them: very dangerous savages indeed. / Employing some semi-civilized aborigines who speak the Japanese a little as guide and collector to climb on the lofty trees, I can easily collect.
>
> I estimate the expenditure on the Formosan expedition £150 (one hundred and fifty pounds) including all the expenses (including the expense on forwarding seeds and specimens to you).
>
> *Yashiroda letter to Johnston, 16 December 1931*

Yashiroda then addresses the costs saying that *I estimate the expenditure on the Formosan expedition £150 (one hundred and fifty pounds) including all the expenses (including the expense on forwarding seeds and specimens to you).*

Lawrence and five other subscribers – the Marquis of Headfort, Lionel de Rothschild, Collingwood Ingram, Sir Stephenson Kent KCB and Hon H. D. McLaren – each paid £25 towards the cost of the expedition to Formosa. Lawrence also arranged that, in return for a share, the Royal Botanic Gardens at Kew would divide and distribute the seeds and plants. The expedition took place during September and October 1932. By November 1932, Kew was sending to the subscribers various lists of the seeds and plants sent by Yashiroda to Lawrence, which he had forwarded to Kew for division and distribution, along with the field notes made by Yashiroda.

Southern Appalachian Expedition, 1933

This expedition was proposed by T.H. Everett, the horticulturist of the New York Botanical Garden, New York in a letter of 25 January 1933 to Thomas Hay, the Superintendent of Hyde Park, London in order to obtain the seeds of plants native to the United States. Everett says that a collecting expedition is planned in the Appalachian Mountains and the Coastal Plain Regions for the autumn of 1933. He adds *I believe that a syndicate of about 20 people, contributing say fifty dollars (at the present exchange rate about fourteen pounds) would be a satisfactory arrangement. We expect, of course, to collect living plants as well as seeds and this sum would take care of the shipping charges.*

Details of the areas to be covered are provided in a note entitled *Portions of the Appalachian Mountains to be covered in proposed collecting trip to be organized by the New York Botanical Garden.* This includes the Allegheny and the Blue Ridge mountains and also includes two areas highlighted as being those with the most interesting material.

---

* Cumberland mountains and Plateau of eastern Kentucky, south-western Virginia west of the Holston Valley, east-central Tennessee, extreme northwestern Georgia, and northeastern Alabama.

---

* Blue Ridge mountains of eastern Tennessee, western North Carolina, northwestern South Carolina and northeastern Georgia, including the Unaka, Iron, Stone, Black, Roan, Nantahala and Smoky mountain regions.

Each of these regions has a flora primarily distinct and unique, the majority of which plants have never been put into cultivation. The most interesting material will be found in the sections marked *

Undated New York Botanical Garden note

The expedition took place as planned and on 28 November 1933, T. H. Everett wrote to the Secretary of the R.H.S. proposing that the RHS should receive all the plants in one consignment and then for the R.H.S. to distribute them to the seventeen shares taken in England. A list of those with shares included Major Lawrence Johnston:

Lt. Colonel C.H. Grey, Hocker Edge Gardens, Cranbrook, Kent, England.	1
Major L. Johnston, Hidcote Manor, Campden, Glos. England.	1
Mr. A.K. Bulley, Ness, Neston, Wirral, Cheshire.	1
The Hon. Henry McLaren, C.B.E., Bodnant, Tal-Y-Cafn, North Wales, Great Britain.	1

Extract from undated list of shares in England

Subsequent letters provided a listing of the plants collected together with information about the conditions for their cultivation:

CONSIGNMENT OF PLANTS COLLECTED BY

THE SOUTHERN APPALACHIAN EXPED. 1933.

DISTRIBUTED FROM WISLEY.

No. I.

Trifolium virginicum shale-barrens, W.Va. A rare and interesting plant for rock-gardens and will almost certainly succeed best in a soil made very open by the addition of liberal quantities of stone chips small cinders or similar material.

Senecio antennariifolius, shale-barrens, W.Va., A matting plant for rock-garden. Culture required probably same as above.

Extract from undated list of plants distributed from R. H. S. Wisley

Lawrence Johnston's plant hunting

An interesting insight into Lawrence's enthusiasm for new plants and also his friendliness with the plant hunter, Collingwood Ingram, with whom he went to South Africa in 1927-8 is shown in a letter from Lawrence in March 1938.

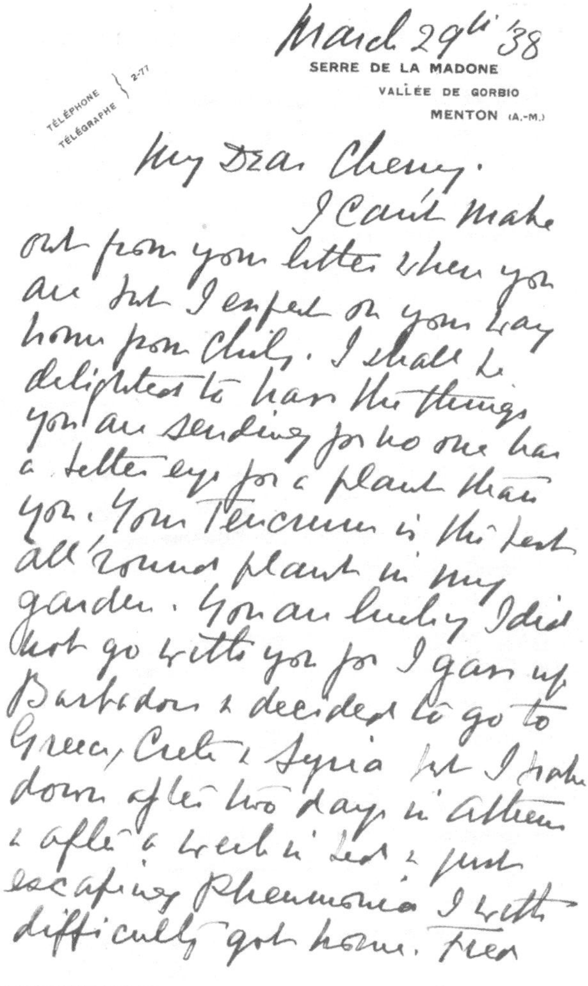
March 29th '38
SERRE DE LA MADONE
VALLÉE DE GORBIO
MENTON (A.-M.)
TÉLÉPHONE TÉLÉGRAPHE 2-77

My Dear Cherry.
I can't make out from your letter where you are but I expect on your way home from Chile. I shall be delighted to have the things you are sending for no one has a better eye for a plant than you. Your Teucrium is the best all round plant in my garden. You are lucky I did not go with you for I gave up Barbados & decided to go to Greece, Crete & Syria but I broke down after two days in Athens & after a week in bed & just escaping Pneumonia I with difficulty got home. Fred

Lawrence Johnston letter to Collingwood Ingram, 29 March 1938

The letter continues saying that Fred Stern, of Highdown, and Lawrence had *decided you are the sort of fellow we want to go and get plants for us, not these high brow botanists with no eyes, only spectacles!* Lawrence goes on to say that *I long to go to Mexico but I am afraid my travelling days are over.* He finishes *My kindest regards to your wife, Yrs. ever, Johnnie.* The warmth of the friendship between Cherry and Johnnie is clear.

During the 1920s and 1930s Lawrence actively sought new plants for his gardens at Hidcote and at Serre de La Madone. In addition to sponsoring plant hunters and making his own expeditions to find plants, he also exchanged plants and seeds with other leading garden owners of the day and in particular with fellow members of the Garden Society.

It is also evident that Lawrence was a generous man who made gifts of plants and seeds to botanical gardens around the world as well as to his friends. Such gifts were made in 1928 and again in 1932 to Kirstenbosch in South Africa, in the 1920s and 1930s to the Cambridge University Botanic Garden and in 1929 and the 1930s to Westonbirt for its arboretum. The plants sent to Kirstenbosch were of particular value as they were recorded as *economic plants* – plants that might be grown commercially. These included *Jasminum grandiflorum* and pelargoniums and three varieties of rose grown for perfume production – Attar of Roses – sent by Lawrence from the south of France which were enthusiastically reported on in the Annual Reports of Kirstenbosch in these years.

> The most important accessions to the collection of economic plants were (1) grafted plants of *Jasminum grandiflorum*, the source of the jasmine perfume for which an enormous demand exists, and (2) plants of the Pelargonium used in the South of France for the production of "oil of Geranium," extensively used in perfumery. Both of these were contributed through the kindness of Major L. Johnstone, to whom hearty thanks are due.
>
> *Director's Report, Kirstenbosch Botanic Gardens, 1928*

Lawrence also produced annual lists of plants and seeds available for exchange which were sent to other garden owners in Britain and around the world.

NINE

HIDCOTE IN ITS HEYDAY

The late 1920s and 1930s saw Hidcote in its heyday. Lawrence Johnston was now in his late 50s, the garden at Hidcote had been completed and he was looking for new and spectacular plants from plant hunters and plant hunting expeditions as well as exchanges of plants and seeds with fellow Garden Society members. At this time the garden at Hidcote was being opened for visitors on two or three days a year – for example, in 1929 it was open with an admission charge of 1s on 5 June from 11am to 7pm for the Queen's Institute of District Nursing.

ENGLISH GARDENS.

LIST OF THOSE OPEN TO THE PUBLIC.

The following gardens will be open in aid of the Queen's Institute and District Nursing during the week ended Saturday, June 8 :—

GARDEN.	COUNTY.	BY COURTESY OF
	WEDNESDAY.	
*Sandringham (11-4.30)	Norfolk	THE KING
Foremarke Hall, Repton	Derbyshire	Major Sir Francis Burdett
Rousdon, Lyme Regis (2-7)	Devon	Mr. P. Vaughan Morgan
Stevenstone. Torrington (2-6)	Devon	Lt.-Colonel B. C. James
Hidcote Manor, Campden	Gloucestershire	Major Johnston
Icomb Place. Stow-on-the-Wold	Gloucestershire	Captain G. H. Simpson-Hayward

The Times, Saturday 1 June 1929

Hidcote was open again the following year on 23 and 30 August for the Children's London Garden Fund. The opening of gardens to the public in aid of the Queen's Institute of District Nursing was the forerunner of today's National Gardens Scheme and was in the 1920s and 1930s supported by the King, who opened Sandringham, and by many of the landed gentry, including members of the Garden Society, who likewise opened their gardens.

A good appreciation of Hidcote in its heyday can be gained from the first published descriptions – two articles in *Country Life* in 1930 by H. Avray Tipping, who was a leading authority on the history, architecture and gardens of English country houses, and one in *The Listener* in 1934 by Russell Page, a landscape artist and garden designer. The first *Country Life* article in February 1930 entitled *Hidcote Manor, Gloucestershire The Seat of Mr. Lawrence Johnston* encapsulated the situation in its subheading

> *A pleasant stone house, of the type that abounds in the Cotswolds. But the gardens are exceptional. They were excellently laid out some twenty years ago, and are admirably maintained.*

He starts by describing the approach to Hidcote as being through ... *solid gates ... hung on to urn-surmounted stone piers such as were usual in Late Stuart days. Here botanic interest could easily be added to picturesque architecture. To the right, between ways into stable yard and kitchen court, simple but shapely buildings were given full value, and between their walls and the spacious gravel area there was room for interesting planting. Thus a Plagianthus [Hoheria] betulinus, with its birch-like foliage and clusters of drooping July blooms, was given wall protection, but quickly shot up above even roof height to enjoy all the breezes of heaven ...*

The second article later the same year, on 23 August, adds further detail *In the forecourt Lilium centifolium was only getting ready to repeat its previous season's bloom heads at a height of eight or nine feet, and Magnolia Delavayi was at the same stage. But the climbing hydrangea (H. Petiolaris) was flowering up to the house eaves, and two twelve-foot high bushes of Plagianthus [Hoheria] Lyalli, pinned against the wall, were hung with thousands of their large pendant white blooms.*

Entrance to Courtyard, Orangery behind left gate post, 1930

From Manor House towards Courtyard entrance – Orangery on right, 1930

The courtyard today is very similar although the Orangery is no longer there as it was removed in the late 1930s.

Orangery interior, Courtyard, 1930

From the Courtyard, Tipping then visited the East Court Garden between the house and the road. He records *Passing through the wicket at the north-east corner of the house, we find the box edged squares of the parterre bedded with lavender blue nemesias and Petunia Silver Lilac, from which rise standards of heliotrope, harmonising with the colour scheme and scenting the air. Great tubs of pink Hydrangea hortensis at the corners complete the arrangement.* Russell Page four years later provided a fuller – and slightly different description:

> *Four beds, edged with the low-growing variegated euonymus, are divided by cobbled paths. These beds are planted out in spring with silvery-mauve violas, and now with pink begonias and Earlham hybrid montbretias. In the centre of each is a standard silvered-leaved centaurea. .... The wall between this garden and the sunken lane has been lowered and made more architectural by two stone urns. Across the road a wide avenue of limes runs up a sloping grass field to a stone group* [a statue of Hercules] *placed on axis with the centre of the*

house. So, by simple means, a sense of space has been given, and a tiny garden been prolonged into the open country.

East Court garden towards Hercules, 1930s

Tipping moved on to the main garden and what he called the great alley running west from beneath the Cedar of Lebanon:

> *A broad central grassway, starting east of the house, runs south of it and up west to a distant pair of wrought-iron gates, and another broad and lengthy way starts out from its high part and runs southward, so that big structural lines, in consecutive and unobstructed sections, give outlook and extent. .... Their sides form part of the screens of the varied adjuncts and dependencies – the small enclosures differently treated and differently furnished, just as in a great house a central gallery, of which the pillared divisions do not hinder the end to end vista, may have, opening from it, a set of cabinets and closets for the display of duly selected and ordered objects of art and vertu.*

The photographs of 1930 show views that will be recognised instantly today. First looking west from the Cedar Lawn:

From the Cedar of Lebanon towards the Stilt Garden 1930

and then looking east from the Gazebos to the Cedar of Lebanon:

From the Gazebos looking through the Red Borders 1930

The planting in the main beds of the Old Garden was described in 1930:

On either side of us we have a broad border to which length of blooming season and continued harmony of tones are given. Greys and pinks predominate, the first by persistent foliage, the second by successional bloom. Tulips begin it in May, and as they go off Eremurus robustus shoots up its tall heads. That best of the sidalceas, Sussex Queen, follows; while a long lasting summer groundwork is afforded by pink snapdragons, above which sway the feathers of Tamarix aestivalis, arch the boughs of Rose Prince de Bulgarie, and twinkle the stars of single pink dahlias.

Old Garden towards Manor House, 1930

Russell Page in 1934 added more information:

Under the cedar tree it starts as a little lawn surrounded by thick squat hedges of box, then it passes between two wide flower beds. .... They continue in autumn a theme of pink established in spring by thousands of pink and red tulips deeply planted, pink lupins, and the tall spires of eremurus. Now later pink star dahlias and snapdragons form a rosy foam from which rise bushes of the pink Tamarix aestivalis and an old-fashioned rose 'Prince de Bulgarie'.

The structure is unchanged today and many of these plants can still be seen there.

From the great alley, Tipping went on to describe the garden to the south of the cedar tree which today is the White Garden:

The White Garden, looking towards the Manor House, 1930

To the south, a narrow flight takes us to the first of the many enclosures. Here phloxes revel, and the mid-September sun shows their brilliant heads with startling vividness against their yew hedge backing. From the yew arch we look back across the paved circle of this gay little retreat and catch glimpses of the grey walled and many-gabled house through the cedar's horizontal boughs;

He later adds further detail:

we see down into the little yew-enclosed phlox garden. The phloxes promise well, but as their time is not yet, gaiety is introduced by the Martagon lilies, white and purple, that stand high among them, while the yew hedges and arches are splashed with the scarlet of Tropaeolum speciosum which ... cannot do better than it does here.

The White Garden, 1930

Tipping then went on to describe what is today the Fuchsia Garden between the Circle and the Bathing Pool Garden:

Next we come to where a small rose garden intervenes between the great alley and the round pool. This rose garden consists, on each side of a central path, of a round bed framed by four segments, and is a show of exquisite roses in exquisite condition, the pink Gruss an Aachen exceeding in floriferousness as she concentrates her endeavours at this season and gives very little of a second flowering. Through this we descend the steps to the great, clear pool so still and ripple-less in the illustration, but liable in this sunny season to become an animated Lido on a small scale, where bathers joy in its limpid freshness.

Bathing Pool Garden towards the Manor House, 1930

Tipping then returns to the great alley to move westwards through the Red Borders:

> *From the rotunda we enter the longest of the alley's sections. Backed by tall hedges, a profuse display of shrub and plant occupies the attention as progress is made along the grassway. Before us rises a flight of steps flanked by pavilions with tall ogival roofs rising up to ball terminals.*

Russell Page described them in slightly more detail a few years later:

> *straight ahead are two long herbaceous borders planted almost entirely in reds and oranges for late summer colour. Japanese maples and the dwarf mountain pine are clumped at intervals, as is Berberis Thunbergii atropurpurea, to subdue too much brightness and offer evidence of skilful and original planting.*

The view to the west shows the Red Borders leading to the steps between the Gazebos with the Stilt Garden beyond.

The Red Borders towards the Stilt Garden 1930

Tipping then moved to the west between the Gazebos to describe the Stilt Garden:

There is the refuge of the pavilions, but besides that we find behind them cubed blocks of tall clipped hornbeams that convince you, by their solidity, that they can stand unmoved the utmost Boreal rage.

Russell Page described this a few years later:

... a pair of charming pavilions. These are small and highly fantastic; brick-built with steeply pitched roofs and interiors tiled and gaily painted, they suggest in miniature all the mannered garden pleasures of the seventeenth century. Beyond them is a slightly raised lawn formally shaped and edged with brick, and on each side are alleys of pleached trees. Then come ... wrought-iron gates leading on to a grassy plateau from which the Cotswold countryside spreads away to the summer blue of distant hills.

Gazebos, Stilt Garden & Heaven's Gate 1930

Heaven's Gate & Stilt Garden, 1930

Tipping also describes the second main axis of the garden – the vista southwards along the Long Walk:

You see before you and below you the second great grass alley. High hornbeam hedges frame it. Other wrought-iron gates end it. It is bent, the fall of the first half being balanced by the equal rise of the second half.

The Long Walk from the south Gazebo, 1930

Russell Page added additional details:

The second main cross vista is axed on the western pavilion. It is simply a wide grass walk with high hornbeam hedges either side. It falls rapidly to a little stream and then climbs again to end far away in a clairvoyee - an iron grille between urn-capped piers. This whole device is in the best classic manner: it gives distance and size to a not enormous garden, and its plainness makes more coherent the intricacies of the various gardens which it separates.

Tipping also referred to the Wilderness – the area of the garden off to the left of the Long Walk on the far side of the stream – by saying:

> *The latest addition, not yet complete, is a considerable patch occupying the end of the rise on the left of the alley as we see it from the pavilion. It had been so neglected that couch grass and thistles were its denizens. Now it has been cleaned, dug and set with all manner of trees and shrubs that yield autumn colour, and as it confronts you from every open spot along the central garden line, it will offer a rich feast of varied tone in the fall of the year.*

Russell Page notes *A new part of the garden, wired for birds, shelter in odd but effective proximity long beds of various Yucca species, berberis, pheasants and flamingos.*

Flamingo pool in the Wilderness, 1930s

Tipping went on to visit the other features of the garden such as the Theatre Lawn:

The tall yew hedge that we have noticed on our right ever since we left the little crossway rotunda is that which forms the southern boundary of a large rectangular and levelled lawn - like so much else at Hidcote, a well separated item. An equal wall of yew bounds it to the north. East, it ends with a row of great beeches, behind which is the house group. West, it is dominated by an outlook platform. To get the lawn level for games much soil must have been removed from its higher end and much of this was used to form the upper end of the slope into a plateau reached by a great set of steps and shaded by a giant beech.

The Theatre Lawn, looking west 1930

Russell Page in describing the same area said:

Behind the house to the North is a big lawn sheltered by a group of beech trees. Fine turf flattened for bowls and a surrounding yew hedge, swinging in a great curve to frame the beech clump, recall the highly sophisticated sylvan splendours of Bramham and Vaux-le-Vicomte.

Both Tipping and Page say the Theatre Lawn was used for games and bowls.

In the area to the north of the Theatre Lawn is the plant house which was noted by Tipping as being:

> *... a large and permanent shelter. It takes the form of a great, moderately heated glasshouse, of which the whole front ...* [is] *removable for the summer season. Thus it ... is an attractive resort at all seasons. ... in the summer, ... the whole looks more like a pergola than a glasshouse. If we begin to chronicle the rarities, we shall never be getting away, and so we will merely note two things. The mop-headed bloom in the right-hand bottom corner is a specimen of Haemanthus Katharinae, some of which Mr Johnston has himself collected on Kilimanjaro. On the other side, rising up to the roof, we see the top boughs of a fine example of Gordonia axillaris.*

Plant House 1930

Russell Page noted that the Theatre Lawn should lead to something interesting:

> *Mr Johnston, delighting, as ever, in giving everything around him its highest decorative expression, has done enchanting things with all the*

plants usually concealed in white-painted greenhouses. His are not white and they are specially designed with a feeling for proportion which makes them worthy of so architectural a garden. In summer, the sides are all removed, exotic climbers ramp half in shade and half in the open sunlight; pots and tubs are hidden by masses of sub-tropical plants; sanded paths, pools for rare water-plants, raised stony beds for succulents, morains for difficult alpines and oleanders set about in painted tubs, all combine to make a very gay museum.

Not only was Hidcote in its heyday in the late 1920s and 1930s, but also Lawrence was one of those who accepted the invitation of the R. H. S. Council to assist in judging the exhibits at the Chelsea Flower Show in 1927.

CHELSEA SHOW.

MAY 25–27, 1927.

Held in the Royal Hospital Gardens, Chelsea.

The following accepted the invitation of the Council to assist in judging the exhibits:

Baker, G. P.
Baker, W. G.
Barker, S.
Barnes, N. F., V.M.H.
Bean, W. J., I.S.O., V.M.H.
Beckett, E., V.M.H.
Bewley, W. F., D.Sc.
Bilney, W. A., J.P., V.M.H.
Bliss, D.
Bowles, E. A., M.A., F.L.S., V.M.H.
Cory, R., F.L.S.
Coutts, J.
Darlington, H. R., M.A., F.L.S.
Findlay, R.
Galsworthy, F.
Gibbs, Hon. Vicary, V.M.H.
Hall, Sir A. D., K.C.B., F.R.S.
Harris, J.
Harrow, R. L., V.M.H.
Humphrey, W.
James, Hon. Robert
Johnston, Major L.
Jordan, F.
Ladds, F.
Lucas, C. J.
McLeod, J. F.
Mason, H. T.
Mawson, E. P.
May, H. B., V.M.H.
Metcalfe, A. W.
Needham, C. W.
Page, Courtney
Pateman, T.
Pilkington, G. L.
Puddle, F. C.
Ramsden, Sir John F., Bt.
Rothschild, Lionel de
Shill, J. E.
Snelling, Miss L.
Stern, F. C.
Stevenson, Thomas
Taylor, T. W.
Usher, W. E.
Wallace, W. E., V.M.H.
Weston, J. G.
White, A. E., V.M.H.
White, A. W.
Wilding, E. H.
Wilson, Gurney, F.L.S.
Yeld, G., M.A., V.M.H.

Judges at the R.H.S. Chelsea Show, 25 – 27 May 1927

Several of his friends from the Garden Society were also judges on this occasion – E. A. Bowles, Reginald Cory, the Hon. Robert James and Lionel de Rothschild. Lawrence was again a judge at the Chelsea Show in 1930.

Lawrence also received Awards of Merit for the following species that he exhibited at various R.H.S. shows:

- Floral Committee, 14 July 1925, Section B, *To Pistorinia intermedia (votes 7 for) from L. Johnston, Esq., Campden. This plant is also known as Cotyledon hispanica. Its stems are straggling, about 1 foot high, succulent, bearing Sedum-like leaves. The flowers are borne in numerous corymbs and are bright yellow, minutely dotted and tipped with red.*

- Floral Committee, 23 October 1929, Section B. *To Gordonia axillaris as a tender flowering shrub (votes unanimous), from Major L. Johnston, Campden, Glos. A very striking plant producing large white flowers like Camellias, and massive, dark green, narrow-oblong leaves. The flowers are axillary, solitary or paired.*

- Floral Committee, 3 June 1930, Section B. *To Halesia carolina var. monticola as a hardy flowering shrub (votes unanimous), from Major L. Johnston, Campden, Glos. A beautiful variety of the Snowdrop Tree. Its flowers are larger than the common form and are slightly flushed with pale pink*

There were other occasions when he exhibited at R.H.S. shows but without receiving an award such as when he exhibited *Ranunculus calandrinoides* (9 February 1926), *Mimulus alpinus, Cyrtanthus 'Pink Beauty', Oxalis spp.* (19 June 1928), *Cyrtanthus obliquus* (3 July 1928), and *Watsonia aletroides, Sparaxis sp.* (25 February 1930).

In addition, Lawrence made donations of plants and seeds to R.H.S. Wisley which included *Cotyledon roseum* (1923), a collection of *Pelargoniums* (1927) and a collection of seeds (1934). The records for the collection of *Pelargoniums* show that he gave some 25 species including *P. Cactus D, P. Scarlet Unique, P. Mrs. Maitland, P. cordatum* as well as some species of *Pelargonium Zonale Cape Town, Pelargonium Zonale Mrs. Zobolisky* and *Pelargonium Zonale minimum.*

Lawrence was also a member of the R.H.S. Lily Committee, set up following an agreement of the R.H.S. Council on 3 November 1931. At the Lily Committee's first meeting two weeks later on 17 November, it was recommended that 12 additional people be added to the Committee including the Hon H. D. McLaren, Hon. Robert James, Major Lawrence Johnston and Mark Fenwick. However, the minutes show that Lawrence only attended two meetings – on 11 January 1932 and 22 November 1932 – and then resigned.

Lawrence's diaries for 1929 and 1932 show that he had a busy social life entertaining friends for lunch and tennis on numerous occasions. The diary entries for the week of 11 to 17 July 1932 provide an example.

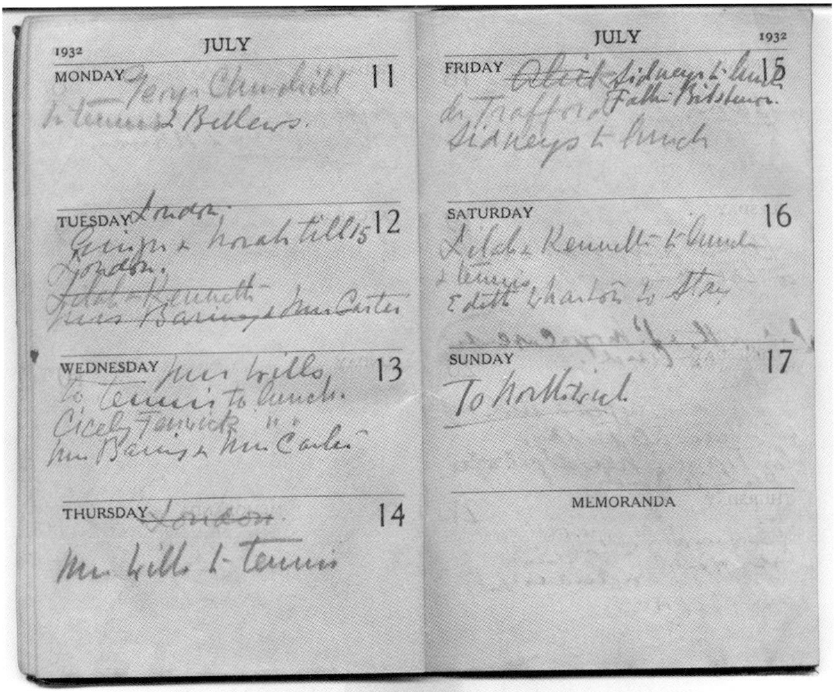

Lawrence Johnston Diary, 11 – 17 July 1932

This shows that on Monday, George Churchill (of Northwick Park) was coming to play tennis as were the Bellews (Lord & Lady Bellew of Barmeath, Co. Louth, Ireland). Ginger and Norah (George & Norah Warre of Villa Roquebrune, Roquebrune Cap Martin, France – along the coast to the west of Menton) were coming on Tuesday to stay until Friday. Mrs. Wills (of Misarden Park, Stroud) was coming to play tennis and have lunch on Wednesday along with Cicely Fenwick (daughter of Mark Fenwick),

Mrs. Baring and Mrs. Carter (socially prominent in London). Thursday again shows Mrs. Wills coming to play tennis whilst on Friday the Sidneys (Col. & Mrs. Henry Sidney of Broadway) are coming to lunch along with Father Bilsborrow (of St. Catharine's Catholic Church, Chipping Campden). Saturday has Lilah & Kenneth (Shennan of Shipton Oliffe Manor, Andoversford) coming for lunch and to play tennis whilst Edith Wharton (the American novelist who lived in France) was coming to stay. The entry on Sunday indicates that Lawrence and Edith were going to visit George Churchill at Northwick Park at Blockley about 10 miles from Hidcote.

Edith Wharton's diary for these years shows that she was quite frequently a visitor to Hidcote and outlined the social activities she and Johnnie (Lawrence Johnston) engaged in together. There are references to being in the garden at Hidcote and then of motoring to Abbotswood to see Mark Fenwick's garden. On another occasion, they visited *Icombe, Mrs Simpson Hayward's place. Incredible beautiful rock, water & wild gardens. Perfect little Elizabethan Manor.* And another time an entry reads *Hidcote. Entire morning in beautiful garden. To lunch the Navarros. In pm we motored to Little Wolford Manor, Broughton, tea with Ld and Lady Saye and Sele.* The following day records that they *started 11 for Nether Lypiatt Manor, lovely little XVII century house on breezy upland in Cotswolds. Lunched there with Mr and Mrs Gordon Woodhouse.* And the next day *Left Hidcote at 11. Lunched Burford. Saw church. At 3.30 Avebury Manor belonging to Mr and Mrs Jenner. Garden and house perfect.* Edith's diary also records similar occasions when she and Lawrence met in France either at Serre de la Madone or at her home in the south of France – the Castel Sainte-Claire at Hyeres on the Mediterranean coast to the west of Cannes.

Lawrence was a good friend of Edith – both were American by birth and both lived in Europe for most of their lives – and he helped Edith create her garden by designing her 'blue garden' at Pavillon Colombe just north of Paris. Edith expresses her disappointment, in a letter on 8 July 1934 to Sibyl Colefax, a notable figure in London society in the first half of the twenty-first century, *that Johnnie is not going to be at Hidcote* and the recent death of her old friend, Lady Antoinette Johnstone, *makes England this summer rather empty and sad.*

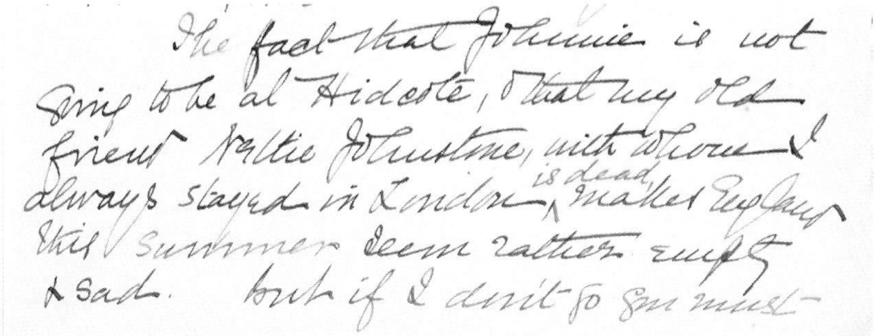

The fact that Johnnie is not going to be at Hidcote, & that my old friend Walter Johnstone, with whom I always stayed in London, is dead, makes England this summer seem rather empty & sad. But if I don't go you must –

Edith Wharton letter to Sibyl Colefax, 8 July 1934

Sibyl Colefax, a notable London socialite, and her husband, Arthur, were also visitors to Hidcote and her photo-albums include photographs annotated *Hidcote Easter 1932* that show the Theatre Lawn and a group of visitors including her husband, Arthur, engaged in a game of bowls. The group, from the left, are Constantine Benson (whose uncle had married Violet, Lady Elcho, the widow of Lord Elcho of Stanway House), Michael Colefax (Sibyl's son), Mrs. Edwina D'Erlanger (the wife of Baron Leo D'Erlanger, a well known hostess in London), Arthur Colefax and Lady Morvyth Benson.

Bowls on the Theatre Lawn, Easter 1932

Another photograph taken on the same occasion shows the two ladies with their feet on the bowls with Hidcote visible in the background.

Lady Morvyth Benson on the left with Mrs. Edwina D'Erlanger,
Theatre Lawn, Hidcote Easter 1932

Another good friend of Lawrence was Norah Lindsay, a keen garden designer, who lived at Sutton Courtenay, Oxfordshire with her children, Peter and Nancy. Lawrence's diaries show that they frequently visited Hidcote and it is known that they also visited Serre de la Madone. Indeed, Lawrence had planned to leave Serre de la Madone to Norah but when she died suddenly in 1948, Lawrence left Serre to Nancy instead.

Further insight into Lawrence's social life can be gained from the entries in the visitor's book at Bodnant which shows that he was staying there for four days from 27 to 30 April 1923 along with Apsley Cherry Garrard, a member of Scott's Antarctic Expedition and author of *The Worst Journey in the World,* and several other keen garden owners who were all members of the Garden Society – C. J. Lucas (Warnham Court, Horsham), Robert James (St. Nicholas, Richmond), William M. Christy (Watergate, Emsworth), Gerard W. E. Loder (Wakehurst) and, faintly in pencil, J. B. Stevenson (Tower Court, Ascot).

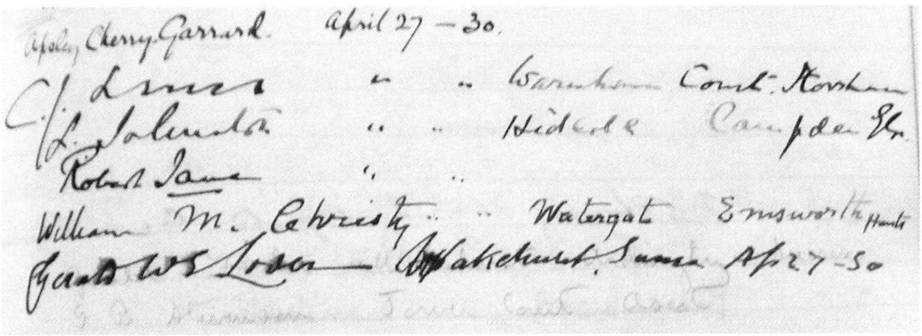

Visitor's Book, Bodnant, 1923

Lawrence stayed again at Bodnant in May 1924, May 1927 and July 1933 on each occasion for four or five days, usually along with other garden owners who were Garden Society members or active in the R.H.S.

Consequently, it can be seen that during the heyday of Hidcote in the late 1920s and the 1930s, Lawrence Johnston was actively engaged socially as well as in activities with the R.H.S. and the Garden Society. He was living a life similar to that of the other landowners who were keen gardeners and spending his winters in the South of France on the Mediterranean Coast at Menton. He was also actively seeking new and attractive plants for his gardens at Hidcote and Serre de la Madone either through sponsoring plant hunters or taking part in plant hunting expeditions. His garden at Hidcote was open for charity on two or three occasions during the year and it was becoming increasingly well known through the articles in *Country Life* and *The Listener.*

TEN

HIDCOTE TO THE NATIONAL TRUST & JOHNSTON'S FINAL YEARS

Lawrence Johnston was living at Serre de la Madone in 1940 when the Italians invaded France. He was then approaching the age of 70.

Lawrence Johnston, Serre de la Madone, May 1940

In June 1940, Lawrence escaped from Serre de la Madone by being taken in an ambulance of the Anglo-American Ambulance Corps of Cannes driven by Comte Yves Rainy du Monceau de Bergendahl, then a boy of about 17, who had escaped from school in Belgium to Menton where his parents were then living. Lawrence was collected in the Val de Gorbio along with twelve dachshunds and three parrots and driven along the Mediterranean coast to Aix-en-Provence, where they stopped overnight and the parrots were left with two ladies. The next day they continued to Le Perthus on the Spanish border where the British Vice Consul for Barcelona, the next Spanish town, was present to help Lawrence cross into Spain.

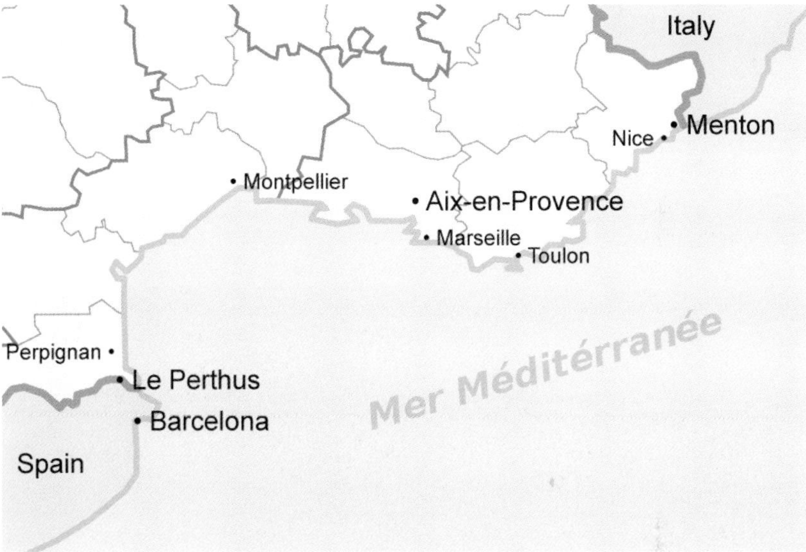

The Mediterranean Coast of France from Menton to Le Perthus

Lawrence then travelled to England by sea and spent the World War II years at Hidcote. In 1943, he participated in a luncheon given by Sibyl Colefax in London. Others at that lunch included several who were then prominent in society such as Osbert Peake, then a Home Office minister, Gaston Palewski, who was General de Gaulle's Directeur du Cabinet, James Hennessy Pope, a biographer and travel writer, David K. E. Bruce who was in the 1960s to be the U.S. Ambassador to the Court of St. James, Thomas H. Eliot, who had recently been a representative for Massachusetts in the US Congress and in 1943 was the director of the British Division, Office of War Information, London and special assistant to the U.S. Ambassador, Edwina Mountbatten, the wife of Louis, Lord Mountbatten, G. M. Young, a historian, and Norah Lindsay, garden designer. James Lees-Milne in his diaries records that *After luncheon, which was delicious, Laurie Johnston took me aside to ask if the National Trust would take over*

Hidcote garden without endowment after the war, when he intended to live in the South of France for good. However, nothing happened following this discussion as it was in the middle of the war years.

Several photographs of Hidcote were taken by *Country Life* in 1943 which give a good indication of how the garden looked in these years and also include parts of the garden that were not included in the photographs taken in 1930 such as the Pillar Garden, the Rock Bank, the Winter Border and the Wilderness.

Entrance to Courtyard, Hidcote Manor 1943

This photograph shows that the Orangery behind the left hand gate pillar that was evident in the 1930 photographs is no longer there. The plants growing alongside the house extend in various places up to and onto the roof.

The view from the Cedar Lawn west to the Stilt Garden, 1943

From the Old Garden towards the Manor House, 1943

The lush growth in the Old Garden is evident.

The Pillar Garden from the Rock Bank, 1943

The Rock Bank, 1943

To the east of the Pillar Garden was the Long Walk with its hedges now well established in a view towards the Gazebos.

The Long Walk looking towards the Gazebo, 1943

Running from east to west by the Gazebos is the Winter Border with Mrs. Winthrop's Garden to the left through the beech hedge.

Winter Border, 1943

From Mrs. Winthrop's Garden there is a view southwards into and through the Wilderness.

The Wilderness, 1943

This photograph was taken from the edge of the former flamingo pool in the Wilderness looking south towards the boundary of the garden.

The following year, Lawrence had a chain survey carried out of Hidcote Manor Garden which provides a valuable record plan of the features of the garden at that time.

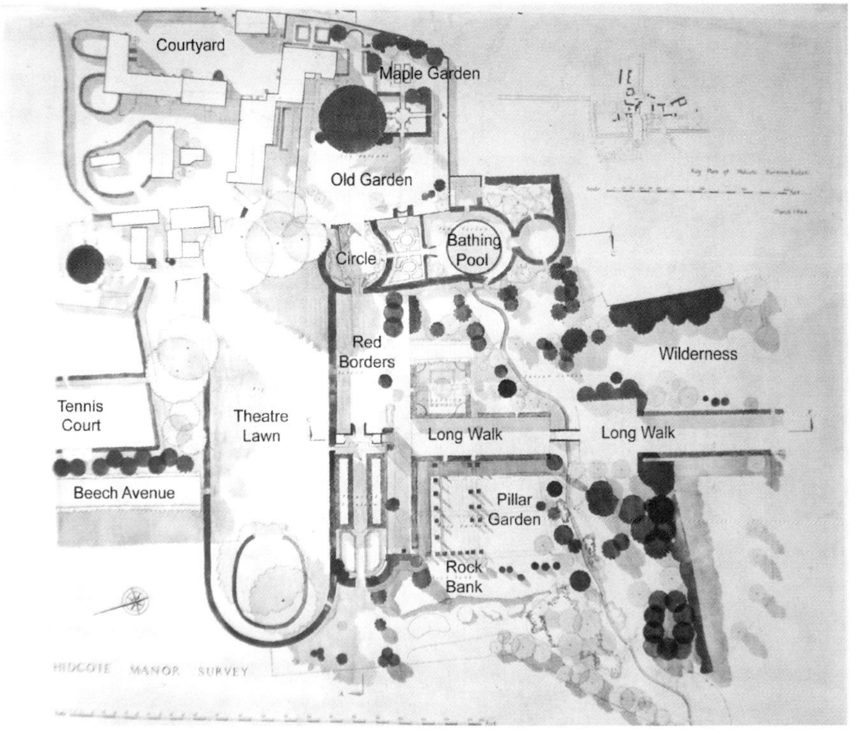

Hidcote Manor Garden Chain Survey, 1944

Following the end of the war, the next development came in a letter from Sibyl Colefax to James Lees-Milne in April 1947 when she says that she had been at Hidcote and that Lawrence wants to give Hidcote to the National Trust now. Her letter starts:

> *I was over at Hidcote with Vivien Leigh Saturday. Laurie Johnston wants to give Hidcote to the N. T. now. So do get him tied up. You see he is not gaga but has no memory. He told me, indeed took me away specially to talk of this.*

She continues to say that, although he had written to Reggie Cooper who has much to do with the National Trust, Lawrence said to Sibyl *Do get*

Lees-Milne to come & settle it with me. Sibyl points out that Lawrence has no relations and adds that *Selling the house would be easy – the farm is a good one – the kitchen garden superb qua stock & condition (it supplied 4 or more huge hospitals during the war) – & 3 miles from Campden, 6 from Broadway & Evesham very easy to make it a very paying concern.*

19 LORD NORTH STREET,
WESTMINSTER.

Sibyl Colefax letter to James Lees-Milne April 1947

This led James Lees-Milne to write on 1 May to Lawrence saying that *Sibyl Colefax had written me a line to say that she has been over to Hidcote, and that you would like to see me and have a talk about the future of Hidcote some time. I need hardly say that I should be delighted to do so.* He went on to say that he was terribly busy for the next few weeks but would probably be staying with his parents at Wickhamford – about 10 miles away from Hidcote – over Whitsun and asking if he could come and visit one afternoon then. Lawrence replied the following day saying *Yes, I have decided to make this over to the National Trust. Anyway at my death. I shall be glad to see you at Whitsuntide.*

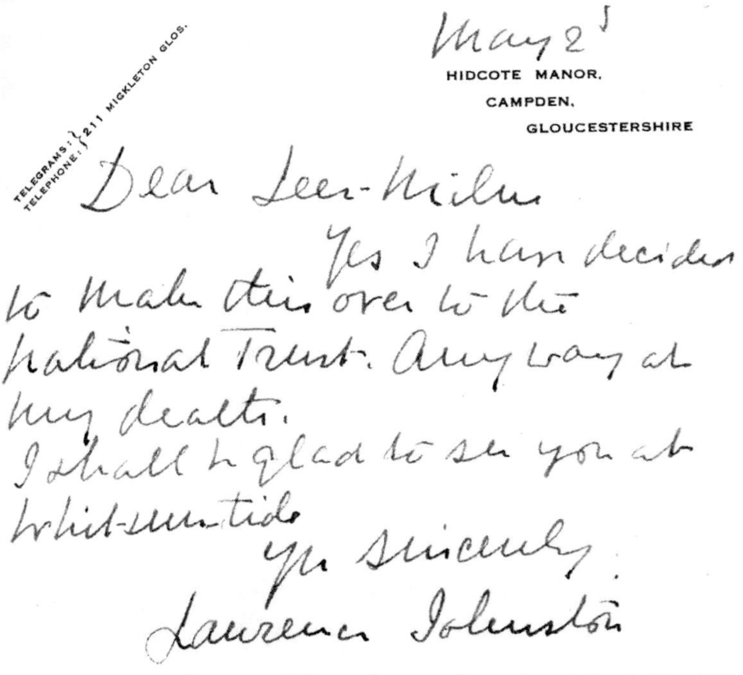
May 2nd

HIDCOTE MANOR,
CAMPDEN,
GLOUCESTERSHIRE

TELEGRAMS: } 211 MICKLETON GLOS.
TELEPHONE: }

Dear Lees-Milne

Yes I have decided to make this over to the National Trust. Anyway at my death.
I shall be glad to see you at Whitsuntide

Yrs sincerely
Lawrence Johnston

Lawrence Johnston letter to James Lees-Milne 2 May 1947

James Lees-Milne visited Hidcote and produced a one page report dated 25 May 1947. This said that Major Johnston offered to leave his property to the Trust by will but wished to know that the Trust will accept it before he made the necessary testamentary depositions. The report noted that the gardens created by Major Johnston over the past 40 years *are by garden experts accounted of national importance and interest. ...*

their layout is such that the visitor is constantly coming upon an unforeseen glade or vista. James Lees-Milne says that *As a specimen of the 20th century garden, this one at Hidcote is fascinating and probably unsurpassed.* His report concludes by noting that *Major Johnston considers that 5 gardeners are the maximum that would be required. He cannot provide any endowment in money, and the Trust would only receive revenue from the letting of the house and the farm rent.*

HIDCOTE MANOR, CHIPPING CAMPDEN, GLOUCESTERSHIRE.

Major Lawrence Johnston, who is elderly and in very poor health, offers to leave his property to the Trust by will. Before he makes the necessary testamentary dispositions he wishes to know that the Trust will accept the devise.

1. The House.

The property consists of the small house, a stone built Cotswold manor, enlarged by Major Johnston and with every modern convenience. It has three or four principal living rooms and approximately 5 bedrooms (excluding servants' rooms). It is a very pleasant little house but not by any means of importance architecturally. All its contents will be left, including linen, silver, crockery. The furniture is pleasant, most of it oak, but again not of any consequence. But the house could be let furnished.

2. The Estate.

The estate consists of 284 acres, comprising one farm, let to a very good and well-known farmer, called Rightson, from Ebrington. He rents it at about £400 p.a., is well contented and tells me he hopes the Trust would allow him to continue. The land surrounds the house on all sides. The situation on the edge of the Cotswolds, overlooking Honeybourne and the Vale of Evesham, is very beautiful indeed. The village, or rather hamlet of Hidcote is included, and is most attractive, grouped on either side of a narrow lane, ending in a cul-de-sac. It consists of the farmhouse, a typical Cotswold building in stone, small village room and about 7 stone built cottages with thatched roofs. The farmhouse is sublet by Rightson, the farm tenant. Most of the cottages are service cottages. The village would however presumably be something of a financial liability.

3. The Gardens.

The raison d'être of the property is of course the gardens, which have been created over the past 40 years by Major Johnston. They are very well known in this part of the country and, indeed, are by garden experts accounted of national importance and interest. In acreage alone they are not large, probably 10-15 acres in all: but their layout is such that the visitor is constantly coming upon an unforeseen glade or vista.

As a specimen of the 20th century garden, this one at Hidcote is fascinating and probably unsurpassed.

Unfortunately, gardens are ephemeral things, unless money is fairly lavishly spent upon them. Major Johnston considers that 5 gardeners are the maximum that would be required. He cannot provide any endowment in money, and the Trust would only receive revenue from the letting of the house and the farm rent.

J. L-M.

James Lees-Milne report, 25 May 1947

The National Trust then began to consider whether to accept Hidcote. The Historic Buildings Committee on 12 June 1947 *decided strongly to recommend acceptance of the devise on account of the gardens, provided (a) adequate endowment was forthcoming and (b) the gardens would be fully maintained and a curator or custodian with proper horticultural qualifications engaged to superintend them in future.* The Finance Committee considered Hidcote on the following day and decided to accept the offer of the Hon. Harold Nicolson to approach the Royal Horticultural Society to see if they could give financial help to the upkeep of the gardens.

James Lees-Milne wrote to Lawrence on 18 June to tell him

> *This is just to tell you that the various Committees of the Trust considered your generous proposal to leave Hidcote to the Trust in your will. First of all, I must tell you that they agreed absolutely unanimously, as indeed I expected, that the Trust would be proud to accept the property if they possibly could, for they attached the very greatest importance to the gardens, but they were not a little worried about the financial position. It was their opinion that without some further revenue than what the farm rents would produce, the Trust could never afford to keep the gardens up in accordance with the standards required. The wages of five gardeners would amount to quite £1,000 a year, quite apart from the other annual expenses needed for the gardens, the cottages and the house, and the need for a Head Gardener or responsible person on the spot to direct operations.*
>
> *The Committee were, however, so reluctant to let the offer pass them by that they asked Harold Nicolson, who is on the Committee, if he would have a private word with Lord Aberconway, Chairman of the Royal Horticultural Society, to see if that body can help us out.*

Lord Aberconway wrote back to Harold Nicolson on 17 June 1947 saying that he had discussed it with the R.H.S. Council and was very sorry to say that it was not in the R.H.S. line of country. He explained that they could not provide funds for a garden that would be rather remote from most of their Fellows and they were not contemplating a horticultural school or college other than what they had already at Wisley.

I was greatly interested in your letter about Hidcote and I discussed the matter generally to-day with the Council of the Royal Horticultural Society. I was very sorry, for many reasons, to have my view confirmed that it was not in our line of country.

We could not provide funds for a garden rather remote from most of our Fellows, nor have we the personnel to supervise the garden and keep it as it should be.

This is sad, but there it is. We have refused other gardens on the same ground. We are not, in fact contemplating a school or college for horticulturists other than we have at Wisley.

Lord Aberconway letter, 17 June 1947

The Trust then made an approach to see whether the Royal Botanic Garden Kew could help but this was also unsuccessful. The National Trust was then uncertain how best to take matters forward although the Chief Agent, Hubert Smith, of the Trust visited Hidcote on 18 August to make a financial report. However, the R.H.S. in November considered the idea of a "Gardens Trust" with Lord Aberconway saying that:

> *He thought that something ought to be done to preserve a few of the outstanding gardens in the Country for the future on the lines of the National Trust and suggested that such a Trust might be established to look after these and other similar gifts. Such a Trust of course would only take over the very best gardens.*

He suggested that this might be under the joint patronage of the National Trust and the R.H.S. and following consultations with the National Trust agreement was reached in February 1948 to set up the *Gardens Section of the National Trust*. This would have as its aim:

> *the preservation of properties with gardens of National importance. Only gardens of special design or historic interest, or gardens having collections of plants or trees of value to the*

Nation either botanically, horticulturally, or scientifically, would be considered.

It would be financed by *The Gardens Fund* to be launched by an appeal for donations, bequests and public subscription. The properties would be administered by *The Gardens Committee* of which half the members would be nominated by the National Trust and half by the R.H.S.

This initiative was announced at the Annual General Meeting of the R.H.S. on 17 February 1948, when Lord Aberconway, the President, said:

> We are undertaking another thing which should be of value in the future, although it will have a small beginning. Under the National Trust schemes, many of the great houses of this country are being preserved for all time for the nation, but up to the present little has been done to maintain gardens in a similar manner, and with the rising costs of upkeep there is a danger that outstanding gardens may deteriorate and be lost unless some action be taken. The National Trust and the Royal Horticultural Society are negotiating an agreement whereby it may be possible to preserve one or two of them. Only gardens of great beauty, outstanding design, or of historic interest would be considered, and those having collections of plants or trees of value to the nation, either botanically, horticulturally, or scientifically.
>
> The National Trust is establishing a Gardens Fund to be maintained by public subscriptions and bequests; these would be used to maintain any gardens taken over by the Trust. The National Trust propose to appoint a Garden Committee, on which half the members will be nominated by the Royal Horticultural Society, to advise on the gardens and the administration of the Gardens Fund. I am sure that Fellows of the Society would welcome such a scheme, and would agree that the Society should make a contribution to the Gardens Fund and appoint members to the Garden Committee. (Applause.)
>
> *Journal of the R.H.S.* 73 *(1948): Proceedings p. xxxviii*

The first meeting of the Gardens Committee on 23 March 1948 was held with Lord Aberconway in the Chair. Members of the committee included Vita Sackville-West, the Earl of Rosse, and Sir Edward Salisbury (Director of Kew) nominated by the Trust and the Hon David Bowes-Lyon and Dr. H. V. Taylor along with the Chairman nominated by the R.H.S. Hidcote was the first item considered under *Gardens to be Considered.* The minutes record that *The Chairman reported that this property has already been offered to the Trust and suggested that it should be one of the first properties to be considered when the Fund has been raised.* The Committee *took note, concurring.*

Shortly afterwards, Lord Crawford, the President of the National Trust – who was also a member of the Garden Society – wrote to Lawrence Johnston. Lawrence replied on 1 April 1948 thanking Lord Crawford for his letter and saying that *It makes it possible for me to make over Hidcote to the National Trust. Naturally after so many years in care of this garden I couldn't give it up entirely and yet as old age comes on I am glad to live in my garden on the Riviera most of the year.*

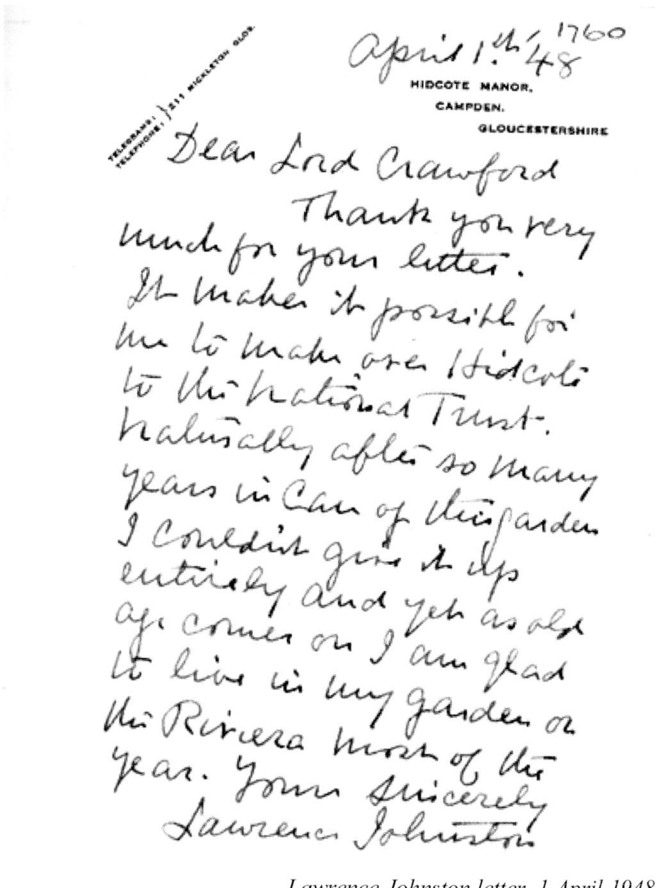

1760

TELEGRAMS: } 211 MICKLETON GLOS.
TELEPHONE:

April 1st '48

HIDCOTE MANOR,
CAMPDEN,
GLOUCESTERSHIRE

Dear Lord Crawford

Thank you very much for your letter. It makes it possible for me to make over Hidcote to the National Trust. Naturally after so many years in care of this garden I couldnt give it up entirely and yet as old age comes on I am glad to live in my garden on the Riviera most of the year.

Yours sincerely
Lawrence Johnston

Lawrence Johnston letter, 1 April 1948

A week later, Lord Aberconway, reported on 6 April to the R.H.S. Council that the Joint Gardens Committee had met and *were of the opinion that the first two gardens, if any, to be taken over by the Trust should be Hidcote and Bramham.* The Gardens Committee met again in April when it was decided to launch an appeal at the end of June to be supported by a letter in *The Times* and a broadcast by Lord Aberconway. The appeal was launched on 28 June 1948

Lawrence makes it clear in his letters that his concern about Hidcote is very much driven by income tax considerations. Thus in a letter on 3 June 1948 to Lord Aberconway he explains that *If I am domiciled here they take all my income in Taxes so I must be domiciled in France where I have a nice house and garden.*

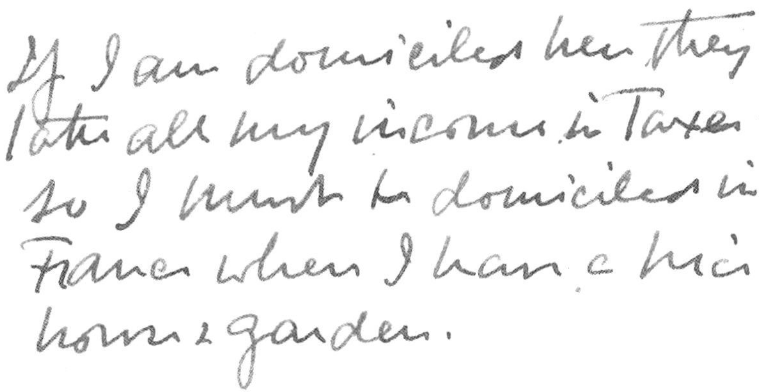
If I am domiciled here they take all my income in Taxes. so I must be domiciled in France where I have a nice house & garden.

Lawrence Johnston letter, 3 June 1948

Lord Aberconway wrote to Lawrence in June saying that the Gardens Committee had decided that *your generous offer of "Hidcote" should be put in the forefront of our Appeal.* Lawrence replied saying that *I am perfectly delighted that you may be able to take over Hidcote. ... Of course, I should like it if I could come back here for short periods in the summer. There is so much I have planted that I should like to see grown. For that privilege I might be able to contribute to the expenses. I should have to consult my lawyer about that. It all depends on whether I should make myself liable for English income tax. I could afford to do it if I did not have to pay income tax.*

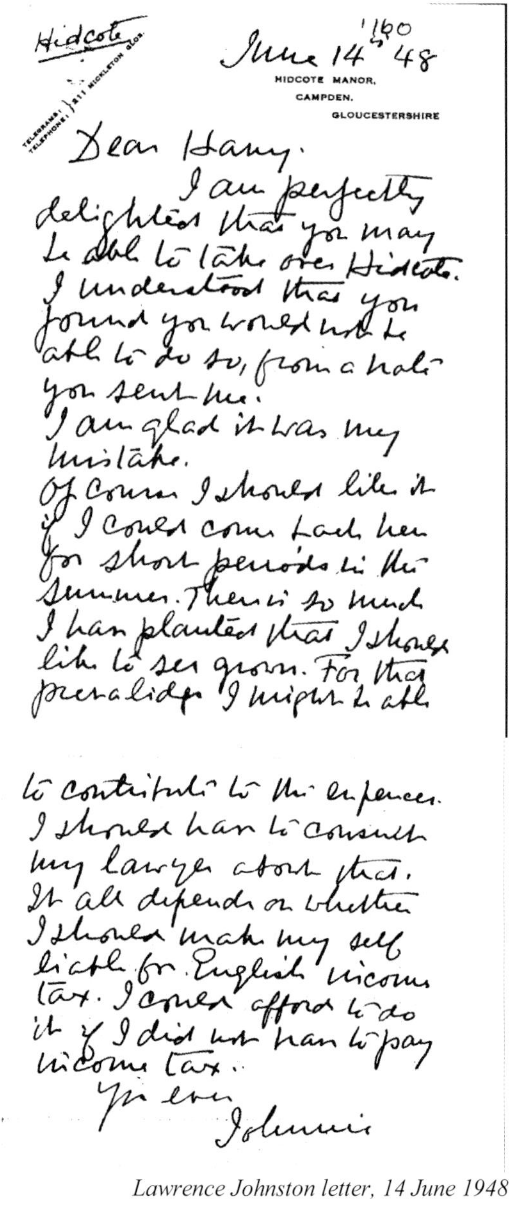

Hidcote

1160

June 14th 48

HIDCOTE MANOR,
CAMPDEN,
GLOUCESTERSHIRE

TELEGRAMS: } 211 MICKLETON GLOS.
TELEPHONE:

Dear Harry.

I am perfectly delighted that you may be able to take over Hidcote. I understood that you found you would not be able to do so, from a note you sent me.

I am glad it was my mistake.

Of course I should like it if I could come back here for short periods in the summer. There is so much I have planted that I should like to see grown. For that privilege I might be able to contribute to the expences. I should have to consult my lawyer about that. It all depends on whether I should make my self liable for English income tax. I could afford to do it if I did not have to pay income tax.

Yrs ever

Johnnie

Lawrence Johnston letter, 14 June 1948

The letter to *The Times* signed by all the members of the Gardens Committee launching the appeal appeared on 28 June 1948.

THE BEAUTIFUL GARDENS OF ENGLAND

TO THE EDITOR OF THE TIMES

Sir,—The many beautiful formal and informal gardens of England are as renowned as the architecturally and historically famous houses to which they are attached. These have had an immense influence on English gardening and indeed on rural life and British character generally, and every one deplores the fact that conditions render it difficult for owners adequately to maintain these national assets. The National Trust is doing valuable work in taking over some of the more historic houses, and now a scheme has been started to save certain of the more important gardens, so that they may be preserved for the nation for all time; monuments, as it were, of English gardening.

The National Trust, in cooperation with the Royal Horticultural Society, has formed a Gardens Committee (of which the signatories are the members) to maintain such gardens and to raise money for the Gardens Fund of the Trust so that this work may be done. Donations to, or covenants for, the Gardens Fund should be sent to the National Trust, 42, Queen Anne's Gate, London, S.W.1, and marked for the "Gardens Fund."

Some outstanding gardens have already been considered, and negotiations for taking these over will be pursued as soon as funds are available. We feel confident that other gardens containing large and valuable collections of plants of scientific or horticultural interest will be offered to the Trust, and it is necessary to make arrangements to maintain these in the national interest. We earnestly hope that all lovers of gardens will respond to our appeal.

Yours faithfully,

ABERCONWAY (Chairman); DAVID BOWES-LYON; ROSSE; V. SACKVILLE-WEST; E. J. SALISBURY; H. V. TAYLOR.

The National Trust, 42, Queen Anne's Gate, S.W.1.

The Times, 28 June 1948

The same issue included a leading article headed *Saving English Gardens* which started by saying that:

To-day the Gardens Committee formed jointly by the National Trust and the Royal Horticultural Society will be launched on its way at a special meeting. A letter on this page asks for help. The aim is to keep up chosen English gardens of special beauty and historical interest, as the Trust already cares for ancient houses.

The following day an article headed *Gardens to be Preserved* described the appeal launched on the previous day and made the point that:

Large funds will be needed merely to preserve the best of our gardens from destruction; even greater sums will be required if they are to be kept as members of the new committee will wish to keep them – not merely static but gaining in beauty with the passing of the years.

It continued *We learn that negotiations have been opened for the acquisition of certain gardens under the new scheme, and among them are those of ... Hidcote Manor, the lovely Cotswold residence of Major Lawrence Johnston.* Two paragraphs described the garden at Hidcote Manor.

> By comparison with Bramham Park it is an intimate garden, although of considerable extent and not lacking in dignity. Small gardens open off a broad central grassway, and the shelter of trimmed yew hedges provides a haven for a host of plants that would be considered dangerously tender for the Gloucestershire climate.
>
> Major Johnston, who brought back many exotic plants from travels oversea, has gathered together at Hidcote rare and beautiful subjects, finding for each a happy setting. The collection of old-fashioned roses and rose species is, perhaps, the finest in the country, and the many magnolias and other flowering trees have now reached mature proportions and would be difficult to surpass in beauty. The formal garden is famous, and the cool, flagged paths of the wild garden lead the visitor through woods where primulas and other shade-loving plants find ideal conditions.

The Times, 29 June 1948

On 12 July, the Gardens Committee noted that:

The Chairman (Lord Aberconway) reported that he had had discussions with Mr. Johnston's solicitors who are anxious that their client should give up his domicile in this country and live in the south of France. The Chairman considered that there was an excellent chance of Mr. Johnston giving the Hidcote Estate to the National Trust now, provided he could reserve a life interest in the house, and that if this were done the Estate would provide some revenue towards the maintenance of the gardens. The Committee decided that Hidcote should have first call upon the resources of the Gardens Fund ...

There was then a flurry of activity as August 1948 was a crucial month because Lawrence had told the National Trust that he would be at Hidcote until 1 September 1948 when he would depart for France. Both the Trust and the R.H.S. considered that the deed transferring Hidcote to the National Trust should be signed by Lawrence before he departed.

A letter, reproduced on the next page, from Lord Crawford sets out the National Trust's understanding that Lawrence will remain supervisor of the gardens for the rest of his life, that the house will not be let and will be at his disposal should he wish to visit Hidcote and that whilst he is at Hidcote, the garden shall not be open more than three days a week, and those days will be at Lawrence's discretion.

Lord Esher and James Lees-Milne arranged to visit Lawrence at Hidcote on 27 August 1948 to obtain his signature on the transfer deed. Lawrence wrote to James Lees-Milne on 17 August saying:

I expect to go to France about the 1st of September. I think the present staff in the garden can carry on. I am head gardener but they have been here so long that they automatically do the work pretty well & if you can spare the time to come here occasionally I believe it would pan out alright. Albert Hawkins is the only real gardener. He is a great plantsman and cultivator but he has not much head for planning ahead. He also runs a big allotment at his home & I am afraid might give that priority if there was no one looking after him. All the same I was away a good deal & things went on all right. The other 3 are very jealous of Albert but they do their work pretty well but I don't think would stand Albert over them. Of course a young head gardener would be best but greatly increases the expense & be certain to alter the character of the garden which is largely a wild garden in a formal setting.

16th August, 1948

Major Johnston,
Hidcote Manor,
Campden,
Gloucestershire.

Lord Esher tells me that he may very shortly be visiting you with a completed Deed of Gift of the Hidcote property from you to the National Trust. I wish to tell you how very greatly I appreciate your generosity in making over these wonderful gardens to the perpetual care of the National Trust.

As to the personal arrangements relating to Hidcote, I am delighted to give you an assurance both on my own behalf, and generally on behalf of the National Trust, that we strongly hope you will remain Controller or Supervisor for the rest of your lifetime of the gardens, which are your sole creation, and that you will, whenever you find it convenient to do so, return to Hidcote. Further, it is clearly understood that during your life the Trust will not let or otherwise use the house and that such furniture as you care to leave there shall be allowed to remain. It naturally follows that whenever you are able to visit Hidcote the house will be at your disposal. Finally, I can assure you that we are entirely agreeable to the point that whilst you are there, the garden shall not be open to the public on more than three days per week and the choice of those days will be at your discretion.

I sincerely trust that these arrangements will be entirely to your satisfaction and would only add that you need have no hesitation whatsoever in accepting this letter as a completely effective and valid assurance of the National Trust's intention.

(Signed) Crawford,

Chairman of Executive Committee.

Lord Crawford letter, 16 August 1948

Lawrence continues in his letter by saying how Peter Lindsay, the son of Norah Lindsay, the garden designer and a good friend of Lawrence, is keen that he should buy a chalet in his valley in the Alps – at Méribel Les Allues in the Savoy region of France – to go to and to send his dogs to in the summer.

Now Peter Lindsay is very keen that I should buy a chalet in his valley in the Alps to go to and to send my dogs to in summer.

Lawrence Johnston letter, 17 August 1948

It is evident that Lawrence was thinking quite a lot about this valley as in a letter a couple of weeks earlier on 31 July to Sibyl Colefax he had said that he remembered going plant hunting there with Mark Fenwick and that it was *one of the loveliest places on Earth.*

and some of the time at Peter Lindsay's valley in the high Alps where I shall take my dogs for the summer. I think I have been in that valley plant hunting with Mark Fenwick & it is one of the loveliest places on Earth.

Lawrence Johnston letter, 31 July 1948

On 27 August 1948, Lawrence in the presence of his lawyer, Mr. Garrett, signed the Conveyance which had been drafted by the National Trust. This states that:

> *In pursuance of his said desire the Donor hereby conveys unto the Trust FIRST ALL THAT messuage or farmhouse known as Hidcote Manor with the cottages buildings yards garden and appurtenances and the several enclosures of arable meadow and pasture land thereto belonging situate in the Parish of Hidcote Bartram or Hidcote Bartrim otherwise Hidcote Barthram in the County of Gloucestershire all which said property contains 261.461 acres or thereabouts and is more particularly described in the First Part of the Schedule hereto ... AND SECONDLY ALL THAT messuage or tenement or farmhouse with the barns stables two cottages outbuildings yards gardens and land forming the site thereof and also all those pieces of land adjoining and thereto all which property formerly formed part of The Hill Farm situate at Hidcote Bartrim aforesaid and now form part of the property known as Hidcote Manor and contains 27.725 acres or thereabouts and is more particularly described in the Second Part of the said Schedule hereto ....*

James Lees-Milne wrote three days later to Lord Aberconway to let him know of the satisfactory transfer of Hidcote to the National Trust.

> You will be glad to hear that Lord Esher's and my expedition to Hidcote on Friday ended smoothly and successfully. Major Johnston signed the Deed of Gift, and seemed perfectly reassured by a letter Lord Crawford wrote to him, which Lord Esher delivered, expressing the Chairman's intentions to allow him complete control of the gardens during his lifetime, and the absolute right to treat the house as his own. It is uncertain precisely when he is going, but probably in two or three weeks' time. Meanwhile Lord Esher feels, and I agree, that it would be a great mistake for any member of the staff to go down to Hidcote again in order to make reports or investigations before Major Johnston leaves. He is easily bewildered and worried.

James Lees-Milne letter, 30 August 1948

The letter went on to report that Nancy Lindsay had been present and that she told Lord Esher and James Lees-Milne that Major Johnston had asked her to supervise the gardens in his absence!

There is another matter that I think you may not be aware of. When we were there, Miss Nancy Lindsay was present; you no doubt know her well. She told Lord Esher and me that Major Johnston had asked her to deputise for him as supervisor of the gardens in his absence. She assumes that this is definitely settled, and it certainly has not occurred to her that the Gardens Committee will even question this arrangement. I think you should know this at once. I have little doubt that Miss Lindsay is a good gardener, and of course she knows the Hidcote garden like the back of her hand, but she is of course very proprietary, and I do not know how well she gets on with the gardeners there.

James Lees-Milne letter, 30 August 1948

Lord Aberconway replied on 3 September saying that he doubted that Miss Nancy Lindsay knew much about gardening and that difficulty would have to be dealt with as it arose.

Miss Nancy Lindsay has a fair knowledge of plants - indeed, as you know, she collected some in Persia, but I doubt that she knows much about gardening.

However, we shall have to deal with that difficulty as it arises.

Lord Aberconway letter, 3 September 1948

Lawrence left for Serre de la Madone on 14 September 1948. He told the National Trust that he would be taking all his dogs, his car and two servants and that he had sent a number of plants from both Hidcote and Kew as well as furniture in a van.

The National Trust now started to manage Hidcote and arranged to have a meeting of the Gardens Committee there at the end of October 1948. The then Secretary of the National Trust, Vice-Admiral Oliver Bevir, wrote to Nancy Lindsay to ask which day would suit her for a meeting at Hidcote when Lord Aberconway, the Secretary and possibly Sir Edward Salisbury, the then Director of Kew, would be present. Nancy Lindsay responded to Lord Esher, instead of Admiral Bevir, in a long nine page handwritten letter saying that she would go and stay the night at Hidcote so either day would be suitable. An extract from her letter shows that she is to be Lawrence's *memory* and *shadow* whilst he is away!

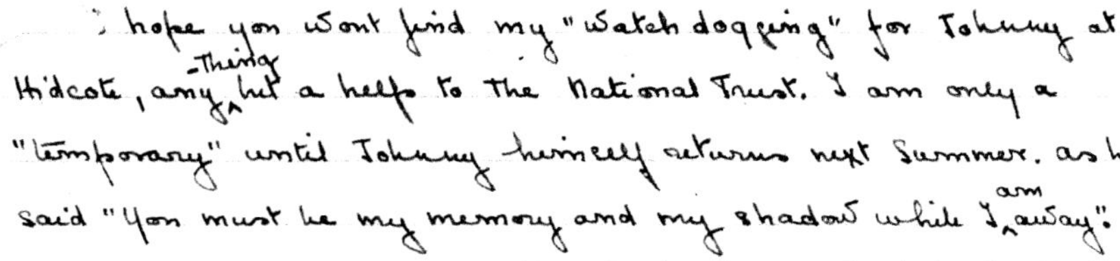
I hope you wont find my "watch dogging" for Johnny at Hidcote, anything but a help to the National Trust. I am only a "temporary" until Johnny himself returns next Summer. as he said "You must be my memory and my shadow while I am away".

Nancy Lindsay letter, undated, October 1948

Interestingly, Nancy had been keeping Lawrence informed as he added a postscript to his letter of 31 October to the Secretary of the National Trust expressing pleasure that Nancy had been appointed *an inspector or supervisor.*

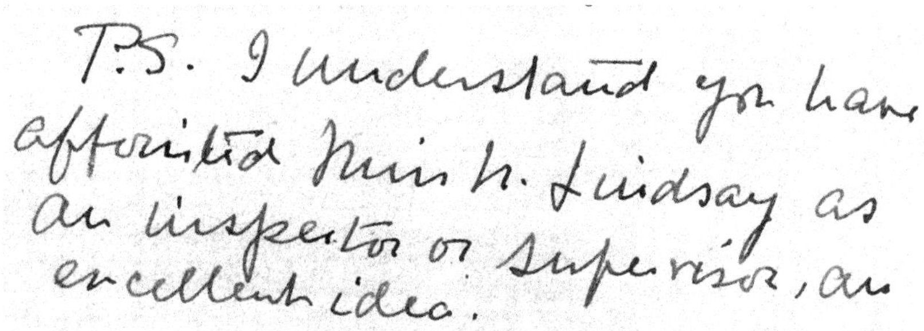
P.S. I understand you have appointed Miss N. Lindsay as an inspector or supervisor, an excellent idea.

Lawrence Johnston letter, 31 October 1948

Following the visit by the Gardens Committee to Hidcote at the end of October, Lord Aberconway wrote on 5 November to Lawrence saying that *It was quite firmly fixed in our minds that we wanted the garden to stay as it was, subject, of course, to such minor changes as you might suggest or approve when you return to England.* The letter went on to say that there were *one or two detailed matters on which we should like to have your views* and said:

> *We thought we ought to enquire whether we could let the Kitchen Garden, leaving in hand the old roses; the Paeonies and the greenhouse and beds pertaining thereto. This, I understand, is what you want, so that there should be no temptation for garden men to work in the Kitchen Garden.*

This letter was sent to Nancy Lindsay for her to forward to Lawrence which she did *with full approval* according to Lord Aberconway's letter sending a copy to Oliver Bevir, the Secretary of the National Trust. There was no direct reply to this letter from Lawrence, possibly because it had been sent via Nancy Lindsay. The next development in regard to the Kitchen Garden was a letter on 12 November from C.D. Inns, Major Johnston's agent in Stratford-on-Avon, to the National Trust Area Agent, Colin Jones, seeking advice following Inns's recent visit to Hidcote when he *raised with the gardeners the possibility of selling surplus plants, and they told me that Miss Lindsay's instructions were that nothing was to be sold, but they could give surplus plants away.* He went on to add that he had subsequently been told that *they have since had instructions from Miss Lindsay that nothing whatsoever is to be sold i.e. no vegetables and in addition no holly. I believe in the past a considerable amount of Holly has been sold to a Birmingham firm for Christmas decorations.* He continued by saying that *As this is contrary to the practice in the Major's time, I feel I must raise the point with you since while I do not wish to question any decisions of Miss Lindsay's, it does seem to me that this is a somewhat foolish policy, and indeed transgressing on the administrative side of the management of the estate.* The matter was referred to Oliver Bevir, the Secretary, who, after consulting the Gardens Committee, wrote on 24 November to Nancy Lindsay to say that *the Trust is very anxious to continue to run the garden exactly as Major Johnston himself did* and consequently the surplus vegetables should be sold and that the practice of selling holly for Christmas decoration should also be continued.

I gather that there is some misunderstanding about the sale of vegetables from the garden at Hidcote and I reported accordingly to the Gardens Committee who told me to write to you about the matter. As you know, the Trust is very anxious to continue to run the gardeb exactly as Major Johnston himself did, and although they know that he does not want the gardeners to spend too much time in the kitchen garden, nevertheless it cannot be allowed to go out of cultivation. Pending possible letting off the garden it would be absurd to grow vegetables and then allow them to rot and the logical thing is to follow what we understand was Major Johnston's practice and to sell the surplus. The Committee are sure that you will agree.

It seems also that a certain amount of holly was trimmed from the trees every year and sold to a Birmingham firm for Christmas decoration. The Committee considered that this should be continued, since presumably the trees must be trimmed, and at about this time of year.

Letter from Oliver Bevir to Nancy Lindsay, 24 November 1948

A lengthy undated four page reply from Nancy Lindsay agreed to the sale of vegetables and of holly. However, Nancy Lindsay did take the opportunity to point out that *Major Johnston had been trying to limit the growing of vegetables and stop the sale of them entirely.* She goes on to add in regard to the vegetable garden that *Before the war it was not that size; it was only ploughed up during the war out of patriotic motives.* She also points out that the practice of making holly wreaths *started because they made a few for the local church and for a few people to whom Major Johnston gave a turkey and a holly-wreath before the War.*

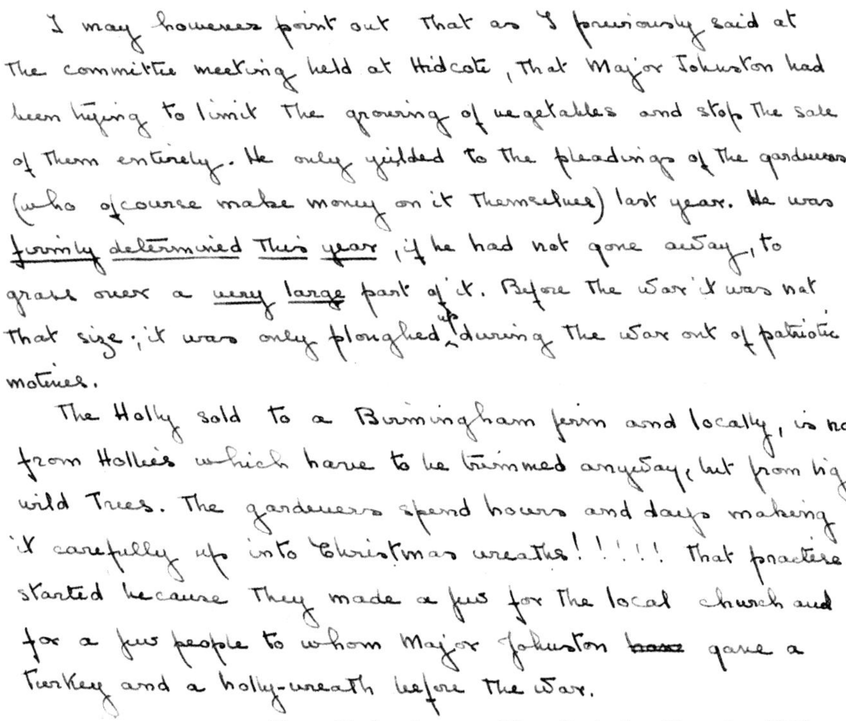

I may however point out that as I previously said at the committee meeting held at Hidcote, that Major Johnston had been trying to limit the growing of vegetables and stop the sale of them entirely. He only yielded to the pleadings of the gardeners (who of course make money on it themselves) last year. He was firmly determined this year, if he had not gone away, to grass over a very large part of it. Before the War it was not that size; it was only ploughed up during the War out of patriotic motives.

The Holly sold to a Birmingham firm and locally, is not from Hollies which have to be trimmed anyway, but from big wild Trees. The gardeners spend hours and days making it carefully up into Christmas wreaths!!!!! That practice started because they made a few for the local church and for a few people to whom Major Johnston ~~has~~ gave a Turkey and a holly-wreath before the War.

Nancy Lindsay letter to Oliver Bevir, late November 1948

Later in the same letter, she says *Before 1939 the Kitchen Garden was not half the size. Vegetables, etc. were never sold. The selling of anything from Hidcote gardens has only grown up during the War.*

Before 1939 the Kitchen garden was not half the size. Vegetables, etc, were *never* sold. The selling of anything from Hidcote gardens has only ~~been~~ grown up during the war.

Nancy Lindsay letter to Oliver Bevir, late November 1948

Nancy Lindsay wrote again to Oliver Bevir in December 1948 to say she had now heard from Lawrence in reply to her letter and that of Oliver Bevir about the Kitchen Garden at Hidcote. She says that *He has thought it over, and now agrees with you that it will be better for the gardeners to keep the Kitchen Garden in "seeming order". Especially as he will be there June-July-August. ... . Major Johnston after all agrees with you not to let off the Kitchen Garden.* This sentence is more an expression of what Nancy thought – the Trust was keeping open the question of letting off the Kitchen Garden until Major Johnston returned to England.

In the event, Lawrence did not return to England in 1949 so in October 1949, the Gardens Committee instructed the Secretary of the National Trust (now J. F. W. Rathbone) *to explore the possibility of setting up a small Local Committee to help to run these gardens.* This led to a Local Committee with Major Kenneth Shennan of Shipton Oliffe Manor, Andoversford as the Chairman together with Mrs. Heather Muir of Kiftsgate, Mr. Joseph de Navarro of Court Farm, Broadway and Miss Nancy Lindsay as members. Nancy Lindsay told the Secretary that she regarded *herself as Major Johnston's "seeing eye"* for Hidcote. Despite misgivings expressed by Mrs. Muir, Major Shennan and the National Trust, it was judged better to include her rather than, as the Secretary noted, *hurt the feelings of this tiresome woman.*

The Local Committee met for the first time on 17 February 1950 at Hidcote. It was advised that its terms of reference might be summarised as follows:

(a) These gardens lack a "boss". Subject to ... the overriding control of the Gardens Committee the Hidcote Committee should, in the absence of Major Johnston, endeavour to make up this deficiency.

(b) The Committee should ... act as a local watch dog for the Trust.

(c) Although the keeping of accounts for these gardens is not the responsibility of the Local Committee the difficult financial position should constantly be kept before them ... and their advice to ... the Gardens Committee should help to reduce the deficit that has to be met by the Gardens Fund.

(d) The Committee is an Advisory and not a Management Committee."

In March 1950 an interesting insight is gained from a letter from the Area Agent to the Secretary of the National Trust in regard to the sale of flowers from Hidcote which notes that Albert Hawkins says that Major Johnston *would not have flowers cut even for the house!*

With regard to the question of selling flowers Hawkins says that he thinks we can let Spry have some (presumably without detriment to the show of flowers in the garden) but adds that it would not be possible when Major Johnston returns as he would not have flowers cut even for the house!

Area Agent to the Secretary, 4 March 1950

Mr. Garrett, Lawrence's solicitor, wrote to the Secretary of the Trust on 23 January 1950 saying that he had just returned from the Continent and had pleasure in sending a cheque for £500 on behalf of Lawrence as a donation to the Gardens Fund. This was the first financial contribution from Lawrence.

Lawrence returned to Hidcote only once, in July 1950, when it became clear that all was not well between him and Albert Hawkins. Indeed, Lawrence wrote to the Secretary saying that *Albert Hawkins seems to be entirely occupied growing fancy geraniums in the greenhouse. They are very lovely but are chiefly indoor plants and don't contribute to the beauty of the garden* and saying that the National Trust could let him go. He concluded by saying that *I am sorry to be such a bother but if I am to live here I must be master in my own garden which I distinctly am not now though I made the whole of it.* This concern about Albert Hawkins passed although it was to surface periodically during this decade.

In August, the Area Agent, Colin Jones, visited Hidcote and wrote to the Secretary that Lawrence seemed quite happy about everything *except the fact that Miss Lindsay was on the Local Committee. He said he hoped we would not bring her to Hidcote as he couldn't bear the woman!* Lawrence returned to France at the end of August. His wish that Nancy Lindsay should have nothing more to do with the house or the garden led to his agent, Captain Inns, seeking confirmation of the instruction given to Albert Hawkins *that Miss Lindsay was to have nothing to do with the gardens or stay in the house.* Lawrence replied by return *I certainly have not asked Miss Lindsay to stay in my house and I should prefer that she should not nor* [sic] *having anything to do with my property.* After discussion with Lord Aberconway as the Chairman of the Gardens Committee, the Secretary wrote to Major Shennan as to whether Miss Lindsay should be retained on the Local Committee. Major Shennan replied that he would be frank and that:

> *1. Major Johnston has taken an aversion to her. Why, I do not know, but, when I saw him last July, it was the one matter on which he seemed to be clear in his mind.*
>
> *2. In spite of her exceptional horticultural knowledge, she is only an encumbrance on the Committee as the mundane and practical points which are so important lie outside her immediate comprehension.*

He concluded by suggesting that professional gardeners should not belong to National Trust Committees and particularly at Hidcote.

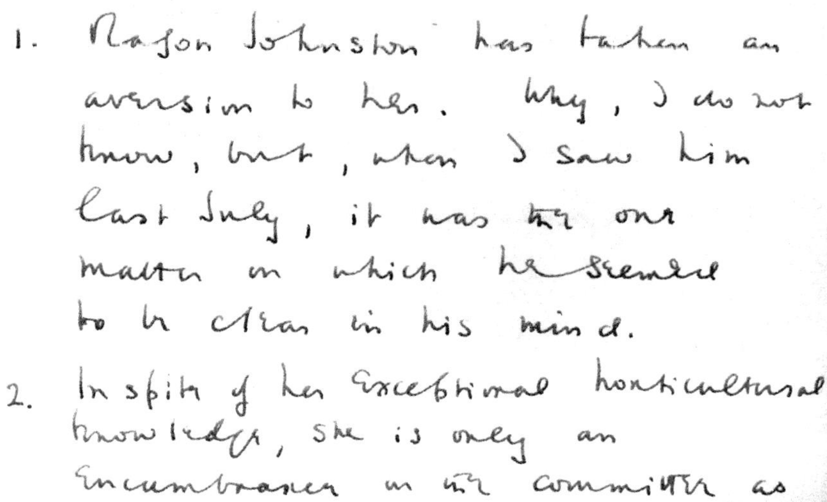
1. Major Johnston has taken an
aversion to her. Why, I do not
know, but, when I saw him
last July, it was the one
matter on which he seemed
to be clear in his mind.
2. In spite of her exceptional horticultural
knowledge, she is only an
encumbrance on the Committee as

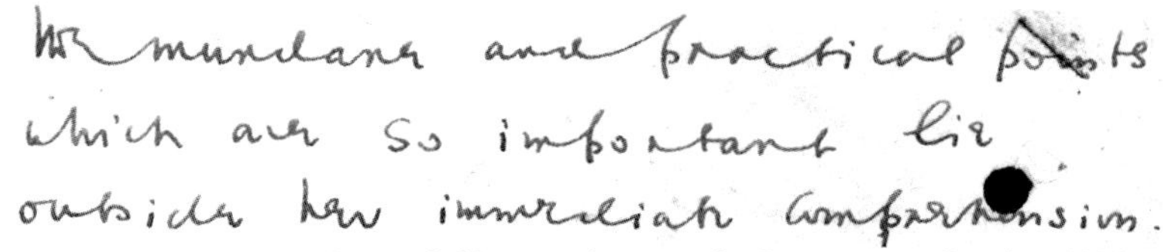
the mundane and practical points which are so important lie outside her immediate comprehension.

Kenneth Shennan letter to the Secretary, 4 October 1950

The Gardens Committee in November instructed the Secretary *to tell Miss Lindsay that she would not be re-elected on the Local Committee for the ensuing year.* He then wrote to tell her that the Gardens Committee had decided that professional gardeners should not be members of National Trust Gardens Committees and she would not be re-elected for the coming year as well as thanking her for her assistance. After some three months, she replied in March 1951 apologising for the delay saying *I can't help being relieved of any share of responsibility of the gardens.*

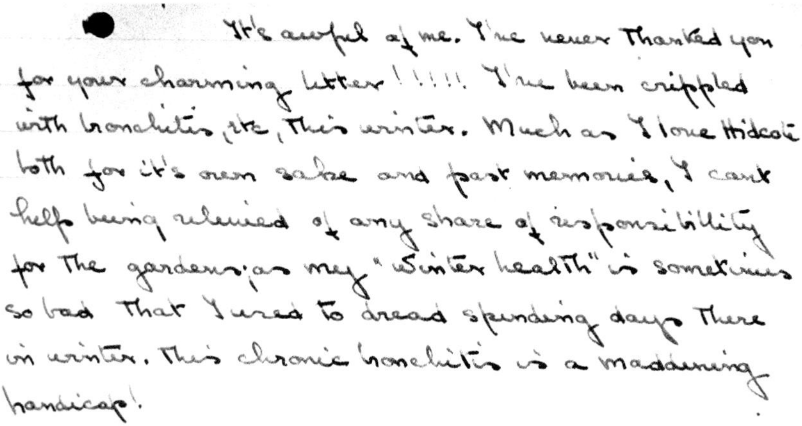
It's awful of me. I've never thanked you for your charming letter!!!!! I've been crippled with bronchitis, etc, this winter. Much as I love Hidcote both for it's own sake and past memories, I can't help being relieved of any share of responsibility for the gardens, as my "winter health" is sometimes so bad that I used to dread spending days there in winter. This chronic bronchitis is a maddening handicap!

Nancy Lindsay letter to the Secretary, March 1951

Concerns again rose about Albert Hawkins and his ability to function as Head Gardener. The Local Committee noted that no thinning or cutting back had been done in the Wilderness for many years and many of the rarer and better species were being smothered by less good species. Lawrence Johnston had put in several shrubs of each species with the intention of cutting out some as they grew. Although additional staff had been taken on to carry out this thinning, the work was not done and the Local Committee had to mark which trees were to be cut out or cut back. Eventually, the situation reached the stage when the Local Committee gave written orders to Albert Hawkins and then checked that these had been carried out.

Lord Aberconway visited Hidcote again in July 1952 when he expressed his satisfaction that, generally speaking, the gardens were in very good order. Instructions given by the Local Committee were to cut back and thin various parts of the garden and to move the *Gruss an Aachen* roses to what was then called the Topiary Garden below the cedar tree – and is today, known as the White Garden.

In October 1953, Major Shennan, the Chairman of the Hidcote Local Committee, pointed out to the National Trust that he had noticed that the tapestries in the hall at Hidcote, which belonged to Major Johnston, were affected by moth. Shennan said that the tapestries are probably of "connoisseur" quality. Arrangements were duly made for three tapestries in the hall and two upstairs to be sent to the Anglo-Persian Carpet Co. for cleaning and treatment for moth. Some of these 'tapestries' were the paintings made on canvas by Lawrence Johnston of the garden and which hung in the bedrooms.

A further donation of £500 on behalf of Lawrence was made by Mr. Garrett to the National Trust Gardens Fund on 20 November 1953. The Secretary wrote to Lawrence saying *I want to thank you for this most generous and welcome donation which will make so much difference to us in the maintenance of Hidcote, a garden, which thanks to you, we are proud to be able to preserve for the benefit of the Nation.*

In the early 1950s the Plant House alongside the Lily Pond was falling into disrepair and consideration was given to whether to repair or demolish it. Although it was initially decided to repair it, the lack of funds led to a decision to demolish it. Sir Edward Salisbury, Director of Kew, visited to identify which plants should be retained in a smaller plant shelter elsewhere in the garden. After a couple of years of debate, the Local Committee decided that it should be pulled down to the approximate height of eight feet only as they considered the building formed part of the plan with the Lily Pond as the centre. Miss Field, the recently appointed National Trust Gardens Adviser, did not agree, nor did the Secretary or the Chief Agent. The Gardens Committee considered the matter in November 1954 and decided on complete demolition. Lord Rosse visited Hidcote four days later and proposed that an extension of the yew hedge would solve the problem of retaining a boundary to that part of the garden. It was eventually agreed to demolish the Plant House whilst leaving the back and ends until the hedge had grown up on the clear understanding that if it looked hideous the back and ends will be pulled down. This was done early in 1955.

Plant House, partially dismantled, 1955

The following year, in July 1956, Major Shennan resigned from the Local Committee saying that:

> *the circumstances have changed considerably since the Committee was appointed in 1949, as the sentimental reasons which then existed for assisting in the preservation of the garden no longer hold good, particularly since the National Trust appointed an expert to advise on the care and well-being of this garden.*

In the same letter to the Secretary he expressed concern about the condition of the house noting that *In the days when Major Johnston was in residence, the house was no less a tribute to his genius than the garden, and it is a tragedy that the National Trust, whose whole purpose is to preserve the beautiful cannot find a way to circumvent the restrictions on its control which arises from the original agreement with its owner.*

In the days when Major Johnston was in residence, the house was no less a tribute to his genius than the garden, and it is a tragedy that the National Trust, whose whole purpose is to preserve the beautiful, cannot find a way to circumvent the restrictions on its control which arise from the original agreement with its owner.

Kenneth Shennan to the Secretary, 9 July 1956

The Gardens Committee agreed to disband the Local Committee at Hidcote and to ask Mrs. Muir to help the Gardens Adviser to look after the garden. The Hon. David Bowes-Lyon, the new Chairman of the Gardens Committee, thanked Major Shennan for *the splendid work you have done for us at Hidcote* and saying that *without your help the situation at Hidcote would have been very grave indeed.* In his reply, Major Shennan ends with a poignant summary *We did our best because of our many nostalgic memories.*

The Secretary wrote to Mrs. Muir saying that the new National Trust Gardens Adviser, Graham Stuart Thomas, should be *supervised and controlled by you.* He added an interesting note of caution that *Without you this plan will not work because, admirable though he is, I should on questions of taste be somewhat chary about giving Mr. Thomas full control.* Mrs. Muir replied saying that

> *I should like to continue helping at Hidcote on the understanding that Mr. Thomas has been notified that we will cooperate on any schemes for the improvement of the garden, as I would not like him to feel that I am interfering in an unofficial capacity.*

She concluded by saying that *As you all know I am very fond of the garden & hope we will be able to improve the borders in the future.*

Later in 1956 following up Kenneth Shennan's letter in July about the condition of the house, and because it was evident that Lawrence was not well enough to return to Hidcote, Mr. Snelling, his accountant and attorney, gave the National Trust permission to store the contents of the house and to let or use the house as they wished. In the event the contents were sold at Hidcote on 31 October/1 November 1956. Mr. Snelling subsequently wrote in February 1957 to the Secretary that the sale had caused him

> *some troubles, as certain persons had suggested that a few personal mementoes should not be sold. However, I got over most of the difficulties but I have to refer to you concerning two questions raised by Miss Lindsay.*

These related to two plants, golden striped Yuccas, which Nancy Lindsay claimed belonged to her and to a curious coloured faience Italian lion from the Plant Shelter which Nancy said Lawrence intended to become her property. After correspondence with Albert Hawkins, the Secretary told Mr. Snelling that these items could be given to Nancy Lindsay.

Progress was also made towards appointing a new Head Gardener as Mr. Snelling had given permission for the National Trust to convert the back part of the house for use as accommodation for a new Head Gardener. An advertisement in the *Gardener's Chronicle* in September 1956 resulted in several applications which were reviewed by Graham Thomas. Although none of these were judged to be of a suitable calibre, in March 1957 Graham Thomas wrote to the Chief Agent saying that he had interviewed an excellent candidate for the post of Head Gardener and recommended that he could be engaged at once. As the Gardens Committee Chairman, David Bowes-Lyon, was away in America and Lord Rosse was in Ireland, it was arranged that Mr. George C. Taylor should together with the Chief Agent interview the candidate in London on 1 May. This took place and led to the post of Head Gardener being offered to Philip M. Knox, aged 31, currently in charge of a woodland garden in Northumberland, subject to a satisfactory reference from his current employer, Colonel Lord Joicey of Etal Manor, Ford, Berwick on Tweed. Lord Joicey replied that Knox had been in charge of the woodland garden for three years, his work had been quite satisfactory and that he had a good knowledge of shrubs and their pruning. Philip Knox was appointed Head Gardener at Hidcote and started there early in June 1957. It is a surprising coincidence that Knox had been gardener at Etal Manor as Lawrence had in 1899 been a farming pupil at New Etal where he had been staying with George Laing, the son of Sir James Laing of Etal Manor. It is very unlikely that Philip Knox will have known that Lawrence used to live in New Etal.

The early years during which the National Trust was looking after Hidcote saw a loss each year which was met from the Gardens Fund. The Annual Reports of the National Trust during the years from 1948 to 1953 show that there was a loss for Hidcote, initially of £400 increasing to about £1,200 a year. The figures for the years after 1953-54 are taken from the National Trust Hidcote files and show an annual deficit of about £2,000.

Year	Property	Expenditure	Income	Deficit
1948-49	Hidcote	£557	£127	£430
1949-50	Hidcote	£2,154	£730	£1,424
1950-51	Hidcote	£2,127	£879	£1,248
1951-52	Hidcote	£2,459	£1,040	£1,419
1952-53	Hidcote	£2,691	£1,397	£2,294
1953-54	Hidcote	£3,587	£1,697	£1,908
1954-55	Hidcote			
1955-56	Hidcote			£2,526
1956-57	Hidcote			£1,572
1957-58	Hidcote			£2,060

Lawrence never returned to Hidcote after his visit in the summer of 1950. In the summers of his later years, he would go and stay at Méribel Les Allues – the valley in the Alps developed by Peter Lindsay. A photograph taken in the summer of 1951 shows Lawrence and his dogs at Meribel – from the left are Will Maker (Lawrence's valet/chauffeur whose name was Gordon Alexander Maker but was known as Will Maker), Lawrence, Freda Bottin (daughter of Fredo Rebuffo), Marie Rebuffo (mother of Freda Bottin and wife of Johnston's butler, Chaiffredo Rebuffo) and Anna, an employee of the property at Meribel).

Lawrence Johnston, second from left, at Meribel, summer 1951

Lawrence Johnston died in Menton aged 86 on 27 April 1958. Will Maker wrote to Nancy Lindsay the same day.

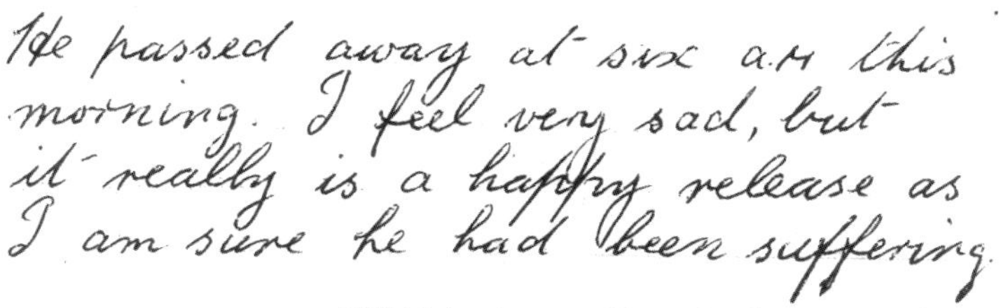
He passed away at six a.m this morning. I feel very sad, but it really is a happy release as I am sure he had been suffering.

Will Maker letter to Nancy Lindsay, 27 April 1958

His body was brought back to England and he was buried in the churchyard of St. Lawrence's Church, Mickleton alongside his mother.

The graves of Gertrude Winthrop (on left) and Lawrence Johnston
Mickleton Churchyard

The inscription on his grave reads *Gifted gardener and horticulturist. Deeply loved by all his friends.*

Inscription on Lawrence Johnston's grave, Mickleton churchyard

In 1964, a plaque was put inside the southern Gazebo at Hidcote with the inscription:

THIS GARDEN
CREATED BY THE GENIUS OF
LAWRENCE JOHNSTON
GIVEN BY HIM TO
THE NATIONAL TRUST IN 1948

Lawrence's contributions to horticulture were recognised by the R.H.S. at the Annual General Meeting on 17 February 1948 when Lord Aberconway, the President of the R.H.S. awarded the Veitch Gold Memorial Medal to him for *his work in connexion* [sic] *with the introduction and cultivation of new plants and for the taste and skill that he has exercised in garden design.* In presenting the award Lord Aberconway said:

The CHAIRMAN: Miss Hudson, I have here the medal given to your cousin, Major Lawrence Johnston. He is a great artist in designing gardens. There has been no more beautiful formal garden laid out since the time of the old Palace of Versailles than that designed on quite a small scale, but with exquisite artistry, by Major Lawrence Johnston at Hidcote. Not only that, but that garden is filled, as the earlier gardens were not, with interesting and beautiful plants, some of which he has himself collected in the mountains of China. No one better deserves the Veitch Memorial Medal than our old friend, Major Lawrence Johnston. Will you give him our very kindest regards and best wishes for his rapid and complete recovery. (Applause.)

Award of Veitch Gold Memorial Medal to Lawrence Johnston, 17 February 1948

Lawrence's will in England had been made on 20 August 1948 before he left Hidcote. This makes it clear that it excludes any property in the Republic of France – for which he made a separate will. When probate for his will in England was obtained on 30 October, the gross value of the estate in England was just over £2,900 and the net value was some £1,660. His principal bequests were largely unrelated to Hidcote and included one of £500 to Mickleton Church to keep the graves of his mother and himself in good order and another of £1,000 to St. Catharine's Church, Chipping Campden, towards the upkeep of the church as well as one of £27 for two Masses to be said each year in perpetuity. Bequests of £1,000 each were made to his cousin, Miss Nan Hudson, his valet Chiaffredo Rebuffo and his chauffeur valet Gordon Alexander Maker. Bequests of £100 each were left to *my head gardener Edward Pearce, my gardener Walter Bennett, my gardener Albert Hawkins* as well as to *Mrs. Dorothy Hughes of Hidcote Bartrim, Miss Nancy Lindsay* of Sutton Courtenay, William Hughes, Miss Annie Bennett, May Bennett and Sidney Nichols of Hidcote Bartrim. The residual estate went to his godson Malcolm Kenneth Shennan of Shipton Oliffe Manor, Andoversford. When probate was obtained on 30 October 1958 this showed that Johnston had been domiciled in France. However, the bequests made under the will in England amounted to over £5,600 and could not be paid from an estate with a net value of £1,660.

Lawrence also left a separate will for his property in France dated 20 August 1948 – the same day as his English will – in which he left Serre de la Madone to Nancy Lindsay. It is probable that this French will of 1948 will have been very similar to an earlier one of 9 November 1946 in which Lawrence named Norah Lindsay as his residual legatee in France and left a little house on the edge of Serre de la Madone in which Chiaffredo Rebuffo and his wife Marie Zelenska lived to them together with some associated land. Norah Lindsay died suddenly on 20 June 1948 at her home in London and it is likely that Lawrence then changed his French will on 20 August 1948 so as to nominate Nancy Lindsay instead of her mother as the residual legatee for his French property.

However, the situation was more complex than might be apparent at first sight from the above two wills, as Lawrence also had funds deposited in the United States amounting to $300,000 (~ £107,000) as well as a small amount (£3,200) in Monte Carlo. A Trust Deed had been drawn up on 23

January 1958 – just a few months before Lawrence died – in which he assigned the sum of £15,000 out of his funds in the Bank of New York to his Trustees – Geoffrey Elmer Garrett, who was Lawrence's solicitor, and Henry Albion Snelling, who was his accountant. This Trust Deed was signed by Henry Snelling on behalf of Lawrence under a power of attorney signed by Lawrence on 26 July 1948, very shortly before Lawrence signed the deed transferring Hidcote to the National Trust. The purpose of the Trust Deed was to pay any French taxes or death duties that might be liable on his French estate and having done this then any balance from the deed was to form part of his residual estate in England.

Following Lawrence's death on 27 April 1958, his executors – who were Geoffrey Garrett and Henry Snelling – started administering his estate. Probate for the English estate was obtained as noted earlier on 30 October 1958. The French and American assets were more complex. It was evident by March 1959, less than a year after Lawrence's death, that Nancy Lindsay was seeking to sell Serre de la Madone. By July 1959, Nancy's solicitor was reporting that he had just returned from Menton when he had along with Garrett attended a meeting with Maitre Michel, Lawrence's lawyer in France. Michel had confirmed that the transfer of Fredo's house had taken place and also said that there could be no further claim for income tax in France. There would, however, be a claim for French death duties against the value of Serre de la Madone which Nancy Lindsay would be liable for. There was, in addition, a danger of a claim being made by the French authorities for French death duties against the assets in America but Nancy would not be liable for this. Three months later, on 5 October, Serre de la Madone was sold to Mr. Baring for the sum of £34,000.

A year later, on 17 November 1960, Nancy's solicitor reports having a *fairly useful talk* at Garrett's office on the previous day. Garrett had compared the effect of Lawrence being regarded as being domiciled in England with being domiciled in France – a domicile in France would result in duties amounting to £89,900 whilst a domicile in England would result in duties amounting to £41,323. Garrett had established that the US authorities would have no objection to a change in Lawrence's domicile and he was going to check with the French authorities who were expected to *probably shrug their shoulders and accept the situation.* This change in domicile was duly made and the probate on the English will was amended on 9 November 1961 to show Lawrence as being domiciled in England.

Progress was also made in respect of the Trust Deed of £15,000. In November 1959, Nancy's solicitor wrote to Humphrey Whitbread, who was a friend and adviser to Nancy, saying that an American lawyer has said that it might be open to Nancy to claim that the property disposed of by the English will passed to her. Nancy's solicitor goes on to say that *We must, I think, all admit that Miss Lindsay has no moral right to the American assets* and then wonders whether they might not put it to Malcolm Shennan's solicitors that Nancy might receive the whole of the £15,000 Trust Deed on condition that Nancy made no claim against the assets in the United States. Some eighteen months later, Nancy's solicitor wrote on 30 March 1961 to say that he had *suggested some time ago to the solicitors acting for Malcolm Shennan that if they agreed to forgo the Assets in Monaco and any share in the Trust Fund of £15,000 in the event of our being unable to justify payment of the whole sum, Miss Lindsay would be prepared to renounce the American Assets.* He adds that Shennan's solicitors have now indicated that they would be prepared to compromise in this way.

The complications, however, continued as it became evident in a letter from Garrett in August 1963 that *the Executor in America will not distribute the funds in his hands unless he can be satisfied that there will never be any claim by the French authorities for death duties. This arises from the fact that by a treaty between France and the U.S.A. French duties can be recovered by the American authorities in the same manner as American taxes. Under the American system the Executor is personally liable in so far as he has not retained funds with which to pay such claims.* By February 1965 it was evident that any possible French claim for death duties would become time barred when a period of 10 years had elapsed. At a meeting on 10 February of Garrett, Snelling and the solicitors for Nancy Lindsay and Malcolm Shennan it was agreed that *the only possible course was to allow things to remain in their present state until the ten year period has elapsed.*

The ten-year period elapsed in 1968 and in the following year, the US funds were released and the various payments made. It was only then that the bequests made under Johnston's will in England could be paid. Having assets in four countries – UK, France, US and Monaco – certainly complicated the settling of Lawrence Johnston's estate. The value of these assets was estimated in 1960 by Garrett as being £1660 in UK, £3500 in France, £107,000 (including the Trust Deed of £15,000) in the US and £3,226 in Monaco – a total of some £115,000 (equivalent to almost £2 million today) which was liable to taxes estimated as being about £65,000 (over £1.1 million today).

EPILOGUE

Following the death of Lawrence Johnston on 27 April 1958, the National Trust continued to look after the garden and estate at Hidcote. The house was let unfurnished for long lets of 7 years at a time, initially bringing in a rent of £250 a year, and these lets continued until 2003. The finances continued to be difficult until the turn of the century when visitor numbers rose to over 100,000 a year and Hidcote became a self-financing property where the money spent by visitors provides the funds to run Hidcote. Two magnificent donations from an anonymous donor – one of £250,000 in 2002 to fund a five year programme and the other of £1.6 million in 2005 to fund a six year programme – both dependent on the Trust raising matching funding – have enabled Hidcote to restore many of the features of the garden to the way that they were in its heyday. The Manor House has been taken back from the long lets so visitors can now walk through the ground floor of the house and enter the garden as Lawrence's visitors would have done and from the semi-circular seat see the superb view along the main axis of the garden through the Old Garden, the Red Borders, the Stilt Garden to Heaven's Gate beyond. The restored features include the Alpine Terrace, the Rock Bank, the Tennis Court, the Plant House, the East Court Garden and the Kitchen Garden. The Conservation Plan for Hidcote, approved by the Gardens Panel on 30 October 2008, recognised that Hidcote's greatest period of significance was the 1930s and consequently to ensure that Hidcote continues to evoke the spirit of its heyday it agreed that the planting policy should follow these principles:

- *Hidcote's historic and aesthetic importance depends equally on superlative design and innovative and excellent planting. However sensitive and imaginative any planting devised by gardeners or gardens advisers might be, the style of planting to the extent possible should focus on what is historically important. It should be as close to the style of Lawrence Johnston, as seen in early photographs of the garden, as we are able to achieve. Johnston's style, as influenced by his circle of friends, should be our touchstone, and only in the absence of clear evidence as to what Johnston intended should it should continue to be presented as refined and enhanced by the National Trust.*

• *It should be visibly of the early 20th Century, not contemporary 21st Century. The balance between keeping to Johnston's original style of planting and keeping the garden alive by reworking the planting is a difficult one to achieve and will demand constant vigilance from all involved to see that, over time, the essential character of the garden does not drift aimlessly. Care needs to be taken to maintain the original scale and proportion intended by Johnston.*

The planting in the garden will thus re-establish Hidcote as it was in its heyday of the 1930s and, in accordance with the goals of the National Trust, ensure its preservation *for ever, for everyone.*

View from Cedar Lawn along the main axis of Hidcote

The future is thus bright for this remarkable garden created by Lawrence Johnston, an outstanding horticulturist and plantsman, who created one of the most beautiful gardens in England.

Lawrence was born in Paris in 1871 to Elliott and Gertrude (née Waterbury) Johnston who had come from wealthy families in Baltimore and New York. His father had served in the U.S. Navy and then fought in the Civil War for the Confederate Forces, losing his left leg, getting an artificial leg and fighting on through the Gettysburg campaign. Lawrence grew up mostly in Europe and then following the divorce of his parents in the early 1880s went briefly to Columbia University, New York before in 1893 coming to England and gaining a degree at the University of Cambridge. He was then a farming pupil at New Etal in Northumberland before becoming a naturalized British citizen in 1900 and serving as a Private in the Imperial Yeomanry in the Boer War during 1900 to 1902. He continued to serve with the Northumberland Hussars, a territorial force, rising through the ranks to Major until he retired in 1922 on reaching the mandatory retirement age. By 1904 he was living at Little Shelford, Cambridgeshire, when he was elected a Fellow of the R.H.S. and started borrowing books on gardening and garden design from the Lindley Library.

Lawrence and his mother bought Hidcote in 1907 and the next seven years were to see the creation of much of the garden. This continued after the Great War ended and the 1920s saw Lawrence become a member of the Garden Society, made up of wealthy landowners who were keen gardeners. He also sponsored and went on plant hunting expeditions throughout the 1920s and early 1930s seeking new and attractive plants for Hidcote. The latter 1920s saw the purchase of Serre de la Madone close to Menton and the creation of a second, very different garden from Hidcote there. Lawrence was, like his mother, a sociable person and entertained many of his friends at Hidcote and at Serre. He never married. He was also generous in the gifts of plants that he made to botanic gardens in Britain and South Africa and his willingness to exchange plants with fellow gardeners.

In 1948 he gave Hidcote to the National Trust. Hidcote was the first garden of national importance to be saved under the Gardens Fund launched by the National Trust and the R.H.S. Lawrence died at Serre in 1958, aged 86, and is buried at Mickleton alongside his mother.

FURTHER READING

Mea Allan, *E. A. Bowles and His Garden at Myddelton House 1865-1954,* London, Faber & Faber, 1973.

Ethne Clarke, *Hidcote The Making of a Garden*, Michael Joseph, 1989; W. W. Norton, 2009.

Daniel D. Hartzler, *A Band Of Brothers: Photographic Epilogue To Marylanders In The Confederacy,* Heritage Books, 2005.

Allyson Hayward, *Norah Lindsay: The Life and Art of a Garden Designer,* Frances Lincoln, 2007.

Collingwood Ingram, *A Garden of Memories,* H. F. & G. Witherby, 1970.

Gertrude Jekyll, *Colour in the Flower Garden,* Country Life & George Newnes Ltd., 1908.

Gertrude Jekyll and Lawrence Weaver, *Gardens for Small Country Houses,* Country Life, 1912.

Gertrude Jekyll, *Home and Garden,* Longmans, Green and Co., 1900.

Gertrude Jekyll, *Wood and Garden,* Longmans, Green and Co., 1901.

Lawrence Johnson [sic], *Some Flowering Plants of Kilimanjaro,* The New Flora and Fauna, No. 5, Vol. 11, October 1929, 11-6.

Norah Lindsay, *Hidcote Manor,* Home and Garden, v.3, April 1948, 46-51.

Thomas H. Mawson, *The Art and Craft of Garden Making,* B. T. Batsford Ltd, 1907, 1912.

Brenda McLean, *A Pioneering Plantsman A.K. Bulley and the Great Plant Hunters,* The Stationery Office,1997.

Brenda McLean, *George Forrest Plant Hunter,* Antique Collectors Club, 2004.

Russell Page, *Hidcote Manor Microcosm,* The Listener, 22 August 1934, 321-3.

Graham S. Pearson, *Hidcote: The Garden and Lawrence Johnston,* National Trust Books, 2007, 2009.

Howard Pease (ed), *The History of the Northumberland (Hussars) Yeomanry 1819 – 1919 with supplement to 1923,* Constable and Co. Ltd., 1924.

Vita Sackville-West, *Hidcote Manor,* J. Royal Hort. Soc., 74, No. 11, November 1949, 476-81.

Karl B. Spurgin, *On Active Service with the Northumberland and Durham Yeomen under Lord Methuen (South Africa, 1900 – 1902),* Walter Scott Publishing Co. Ltd., 1902.

Avray Tipping, *Hidcote Manor, Gloucestershire The Seat of Mr. Lawrence Johnston,* Country Life, 22 February 1930, 286-94.

Avray Tipping, *Early Summer at Hidcote Manor,* Country Life, 23 August 1930, 231-3.

Fred Whitsey and Tony Lord, *The Garden at Hidcote,* Frances Lincoln, 2007.

ACKNOWLEDGEMENTS

I am especially grateful to Mike Beeston, Property Manager, to Glyn Jones, Head Gardener, to John Lumby, Retail Manager, and all the team at Hidcote for their encouragement and assistance in the writing of this book. I would also like to thank the many individuals in both this country and abroad who have provided information and helped me in my search for information about Lawrence Johnston and Hidcote: Lady Ann Aberconway, John Balmford, Gwen Bell, Marguerite Bell, Audrey Bowker, Father John Brennan, Ewen Buchanan, Derek C. Bull, Dick Carter, Ethne Clarke, Jennifer Comins, Stéphane Constantin, Judith Ellis, Brenda Faulkner, James Finlay, Margaret Fisher, Henri Garrigue, Julian Gibbs, Hon. Jane Glennie, Alan Hall, Dorothy Hart, Daniel Hartzler, Allyson Hayward, Ann Hettich, Christine Hiskey, Carol Jackson, Bernard Jeanty, Andrew Joicey, Richard Kernick, Richard Kettle, David Kingsmill, Linda Knight, Sarah Legg, Michael Likierman, Nancy McLaren, Brenda McLean, M. K. Miles, Comte Yves Rainy du Monceau de Bergendal, Hon. Gerard Noel, Edwin Nutbourne, David Owen, Hazel Payne, Douglas Pearson, Linda Phelpstead, Ernest Pollard, Rosemary Powell, Maurice Ribbans, Ros Roberts, Lily Rogers, Martin Smith, Jack Sully, John Tankard, Judith Tankard, Tess Taylor, Gordon Thompson, Arthur Trainor, James M. Waterbury, William Waterfield, Samuel Whitbread, Doris Williams, Edward Wilson, Chris Wynn, Peter Yardley and Wiliam Lee Younger.

Many thanks go also to Melanie Aspey, Director of the Rothschild Institute, Denis Edelin and Rodney Ross of the United States National Archives and Record Administration (NARA) in Washington, D.C, Brett Elliott, Archivist and Librarian, Charlotte Brooks and Lucy Waitt of the Royal Horticultural Society, Roberta Goldwater of The Soldier's Life, The Northumberland Hussars, Discovery Museum, Jackie Graziano of the Westchester County Archives, Mary Hodgson lately of the Legal Section of the National Trust, Elliott O'Neill of the Maryland Historical Society, Dan Monahan of the Green Mount Cemetery, Baltimore, Sindy Pagan of the Woodlawn Cemetery, New York City, Leonie Paterson, Archives Librarian, and Graham Hardy of the Royal Botanic Garden Edinburgh, Jacqueline Cox of the University of Cambridge Archives, Juliet Day of the Cambridge University Botanic Garden, Jonathan Smith of Trinity College, Cambridge, Iain Shaw and Sarah Compton of the Records Section of the National Trust and Vicky Skeet of the National Trust Photographic Library. I would also like to thank the Archives de Paris, the Bodleian Library, the Chipping Campden Library, the Campden and District Historical Association (CADHAS), the Evesham Library, the Gloucestershire Archives, the Mairies of Ferney, Menton and Nice, Her Majesty's Court Service, the National Archives, the Oxford County Studies Library, the Seattle Public Library, and the United States National Archives and Record Administration (NARA) in Morrow, Georgia and Seattle, Washington.

PICTURE CREDITS

Hidcote Books are committed to recognising the intellectual property rights of others. We have therefore taken all reasonable efforts to ensure that the reproduction of all content on these pages is done with the full consent of copyright owners. If you are aware of any unintentional omissions please contact us directly so that any necessary corrections can be made for future editions.

Lady Ann Aberconway: 148U, 148L, 181, 183, 184U, 184L, 209
Mary Anderson, *A Few More Memories*: 65
Bodleian Library, University of Oxford/M.S. Photog.b.10, fol 132, fol. 131: 207, 208
Freda Bottin: 128L, 246
© Derek C. Bull: 57, 74, 111, 253
Syndics of Cambridge University: 38, 39
Cambridge University Botanic Garden: 150U, 150L
Illustrated Chronicle, Newcastle, 27 October 1914: 94U, 94L, 95
Columbia University, Alumni Directory 1932, p.449: 35U, 35L
© Country Life: 90, 105L, 106, 110, 114, 115L, 189U, 189L, 190, 192U, 192L, 193, 194, 195, 196, 197, 198U, 198L, 199, 201, 202, 212, 213U, 213L, 214U, 214L, 215U, 215L, 216
Courtesy of Gloucestershire Archives/Bruton Knowles & Co, D2299/1021: 68, 88, 89, 120, 121
Courtesy of Gloucestershire Archives/CADHAS Jesse Taylor photographs: 67, 70U, 71, 80L, 82L, 84L, 105U, 119L, 125L, 125R, 126
Courtesy of Gloucestershire Archives/ Mickleton Parish Records, P216: 138
Courtesy of Gloucestershire Archives/National Trust, D2784/3: 59, 60
Courtesy of Gloucestershire Archives/New and Saunders, D5163/1/39b: 61
© Miles Hadfield: 200
Daniel D. Hartzler, *A Band of Brothers: Photographic Epilogue To Marylanders In The Confederacy* 13
Collingwood Ingram, courtesy of Ernest Pollard: 154, 155, 156, 157, 159, 160, 161, 162, 163, 185
Kirstenbosch Annual Report, 1928, p.5: 186
LancasterHistory.org 89.82_Harriet_Lane,_ca_1857-1861.jpg: 14
A. H. Lealand: 78
Sarah Legg: 108
Library of Congress Prints and Photographs Division http://www.loc.gov/pictures/ item/93508122/: 20
Reproduced from the London Gazette, Issue No. 27160, 2 February 1900. © Crown Copyright: 48
Reproduced from the London Gazette, Issue No. 27799, 30 May 1905. © Crown Copyright: 53
Maryland State Archives, am565--180.pdf: 6
Thomas H. Mawson: 72, 115UR

Mickleton Women's Institute: 36, back cover
© Dan Monahan: 24, 44
© Comte Yves Rainy du Monceau de Bergendal: 210
The National Archives, WO 374/37873: 101
National Archives and Records Administration, Confederate Records M331, Roll 142: 8, 11, 12
National Archives, Center for Legislative Archives: 30U, 30L
National Atlas of the United States, 3 Feb 2010, http://nationalatlas.gov: 5, 16, 26
© National Trust, Legal Section: 69,102
© National Trust/Hidcote: Front cover L, 66U, 66L, 70L, 77, 79, 80U, 81, 82U, 84U, 98, 99, 104, 109U, 112U, 113, 116U, 119U, 167, 205, 217
Naval History & Heritage Command: 7
New York County Surrogate's Court: 42, 142
Northumberland Hussars, Regimental Archives: 50, 55, 56, 92
Reproduced from the 1885 Ordnance Survey map © Crown Copyright: 60
Images and Voices, Oxfordshire County Council, Packer Collection: 123
© Graham S. Pearson: 1, 19, 21, 38, 40, 46, 47, 75, 107, 109L, 112L, 115UL, 116L, 124, 128U, 131L, 140, 247L, 248U,
Hazel Payne: Front cover UR, ii
Virginia Peacock *Famous American Belles of the Nineteenth Century,* p.169: 15
Photo Precision Ltd: 242
Peter Pritchard: 83
The Rothschild Archive, London: 118, 146, 180, 182
Reproduced by kind permission of Royal Botanic Garden Edinburgh: 172,173, 179
Royal Horticultural Society: 51, 85U, 143, 203, 223, 248
Serre de la Madone: 131U, 132
Southill Park Archives: 133U, 133L, 134U, 134L, 135U, 135L, 136U, 136L, 137U, 137L, 247U
St. Peter's Church, Westchester: 22
© John Tankard: 28
The Times, London, 28 June 1948: 227
© The Times, London, 29 June 1948: 228
© The Times, London, 10 July 1906: 55
© The Times, London, 1 June 1929: 187
© The Times, London, 7 June 1911: 85L
The Master and Fellows of Trinity College Cambridge: 37
Library of Virginia, Richmond: 28, 43
Derek R. Walker, http://www.allatsea.co.za/ unioncastle/carnarvon.htm: 152
Westchester County Archives: 18
Edith Wharton: 207
Mrs. D. M. Williams: 191

INDEX

Page numbers in **bold** denote illustrations